Biotechnology Ethics, Risks and Code of Conduct

Biotechnology Ethics, Risks and Code of Conduct

P. R. Gilbert

ANMOL PUBLICATIONS PVT. LTD.
NEW DELHI - 110 002 (INDIA)

ANMOL PUBLICATIONS PVT. LTD.

H.O.: 4374/4B, Ansari Road, Darya Ganj
New Delhi-110 002 (India)
Ph.: 23278000, 23261597

B.O.: 1015, Ist Main Road, BSK IIIrd Stage
IIIrd Phase, IIIrd Block
Bangalore - 560 085 (Karnataka)
Ph: 080-41723429, Tel/Fax: 080-2672 3604
Visit us at: www.anmolpublications.com

Biotechnology Ethics, Risks and Code of Conduct

First Published, 2008

ISBN 978-81-261-3278-2

PRINTED IN INDIA

Printed at Mehra Offset Press, Delhi.

Contents

Contents

Preface

This book entitled "Biotech Ethics, Risks and Code of Conduct" is designed to acquaint readers with the modern *biotechnology* ethics, risks and code of conduct. Around the world, different countries have different rules relating to cloning – some don't allow reproductive cloning, but do allow therapeutic cloning, while other countries allow both types. In South Africa currently, the Human Tissue Act prohibits both therapeutic and reproductive cloning of humans. This situation will change when chapters 6 and 8 of the National Health Act are promulgated—and therapeutic cloning will then be allowed under strict conditions, requiring Ministerial permission, but reproductive cloning on humans will remain strictly prohibited.

In China, there are no laws against cloning either. However, with the huge population problem and the policy of only one child per family, there is no interest in reproductive cloning. On the other hand, there is huge interest in therapeutic cloning and, as a result, there is a lot of it on the go in China. Some Chinese cultures believe that humans only become people when they participate in society – so, according to these cultures, embryos and foetuses are not considered to be human beings, thereby eliminating any ethical problems surrounding the creation and destruction of embryos to get the required stem cells. The Chinese government is funding extensive research and drawing back many Chinese scientists from overseas, to undertake work they would not be allowed to do elsewhere. With easy access and no limits on obtaining the embryonic material they require, it is expected that Chinese scientists will race ahead of the rest of the world in therapeutic cloning technologies. In the UK there are clear rules banning reproductive cloning, but scientists in both the UK and Israel are allowed to generate new embryonic cell lines for therapeutic research. The law in Germany bans the extraction of stem cells from human embryos for research within the country, but in 2002 a new law was passed allowing some human embryonic stem cells to be imported. This means that German scientists are allowed to undertake work on embryonic stem cells if they originate from outside Germany. In the United States there is no public funding for research on embryonic stem cells.

Therefore, although there is no law against therapeutic cloning in the US, there is almost no public research happening due to the lack of money and access to the embryonic material. Simultaneously, an announcement was recently made stating that US public funds (about US$1,4 million, about R11,2 million) will be provided for studies using adult stem cells instead of embryonic stem cells. These different rules for different countries mean that if a scientist is banned from undertaking cloning in his/her home country, he/she can simply move to a country where it is allowed. To prevent this from happening, the US and about 30 other countries want a global ban on all forms of cloning. However, many countries don't agree and thus no progress is being made on implementing this ban. We could choose to have changes made to us, but we might also be making the choice for our children if the changes are carried through to the germline. Do we have that right, and how far should we take our ability? Conversely, is it responsible and ethically acceptable to leave the potentials of our children to the chance effects of the "genetic lottery", if we obtain the technological capacity to make positive changes? If genetic engineering became the way of the future, would people whose parents could not afford to genetically 'modify' them while still in an embryo, have a chance of achieving with high standards compared to the people who were 'modified' to be perfect? Is it ethical to experiment on embryos that have yet to be born? How would genetic engineering be used to revolutionize warfare? Who decides which changes will be made? Major ethical considerations include the aforementioned. For example, today, women are faced with a rapidly expanding array of reproductive technologies. Developed by private biotechnology companies and marketed to fertility clinics, these new options have been presented to women as an issue of "choice"—supposedly providing them greater control over the process and outcome of their pregnancies.

The Council for Responsible Genetics unequivocally supports a woman's right to make her own reproductive decisions. However, we oppose the utilization of human eggs and embryos for experimental manipulations and as items of commerce because of the potential for eugenic applications and health risks to women and their offspring. The realm of assisted reproduction has become a multi-billion dollar industry, visible in the increasing availability of vitro fertilization, prenatal genetic diagnosis, and chemical and chromosomal testing of the fetus. So far, these experimental procedures have not been closely regulated. Part of this is because the reach of federal oversight extends exclusively to publicly- funded research, leaving private sector activity largely unregulated. Many are resistant to any regulations over the fertility industry due to the continuing political residue of the abortion debate. Women in the United States fought hard to have courts recognize a women's right to choose abortion, and are wary of opening the door to government regulation of anything to do with women's eggs, embryos or fetuses. The word "choice" has powerful connotations and women's groups themselves are not eager to restrict its power. The industry is aware of this and has capitalized on the situation by supposedly offering the new genetic technologies as "choices." The unique role of women in reproduction puts them on the front line of biotechnological experimentation, and as such, women have the potential to play a leading role in determining the direction and scope of these developments. CRG works, in partnership with a number of women's health groups, to reshape the discussion about genetic technologies in reproduction and to equip women with the information necessary to lead the process. Our goal is to shift debates over genetic technologies away from abortion politics and into more effective and productive discussions about the integrity of reproduction and the control of women's bodies. Through education, networking, outreach, and activism women can direct their own reproductive futures.

This book consists of—Biotechnology Ethics: Issues and Views; Biotechnology: Risks and Safety Options; Biotechnology Policy, Guidelines and Statutes; and Regulation of Biotechnology: Model Framework etc. besides a large body of list of acronyms, glossary of relevant terms, extensive bibliography, list of websites and links, for further referencing and research.

—Editor

List of Acronyms

ACMG	—	American College of Medical Genet.
AEC	—	Atomic Energy Commission
AES	—	American Electrophoresis Society
AFMR	—	American Federation of Medical Research
AFIP/ARP	—	Armed Forces Inst. of Pathol./Am. Registry of Pathol.
AGSG	—	Alliance of Genetic Support Groups (now Genetic Alliance)
AGT	—	Association of Genetic Technologists
AHA	—	American Heart Association
AIBS	—	Am. Inst. of Biol. Soc.
AMIA	—	American Medical Informatics Association
BTCI	—	BioPharmaceutical Technology Center Institute
CF	—	Cystic Fibrosis
CFF	—	Cystic Fibrosis Foundation
CHH	—	Cartilage-hair Hypoplasia
CHI	—	Cambridge Healthtech Inst.
CHOP	—	Children's Hospital of Philadelphia
CIMB	—	Center for International Meeting on Biology

CIOMS	—	Council for International Organizations of Medical Sciences
CIRB	—	Colorado Institute for Research in Biotechnology
CLMA	—	Clinical Laboratory Management Association
CMT	—	Charcot-Marie-Tooth
EMBO	—	European Molecular Biology Organisation
EMG	—	Encyclopedia of the Mouse Genome
EMS	—	Environmental Mutagen Society
EORTC	—	European Organization for Research and Treatment of Cancer
ERDA	—	Energy Research and Development Administration
ERI	—	Eleanor Roosevelt Institute
ES	—	Embryonic Stem
GBASE	—	Genome Database of the Mouse
GBR	—	Global Business Research
GDB	—	Genome Database
GMD	—	Genomic Map Design
GPI	—	Genetics and Public Issues Program
GRAIL	—	Gene Recognition and Analysis Internet Link
GRC	—	Gordon Res. Conf.
HMDP	—	Homology Database
ICGEB	—	International Centre for Genetic Engineering and Biotechnology
ICHG	—	International Congress of Human Genetics

ICPEMC	—	International Commission on Protection Against Environmental Mutagens and Carcinogens
ICRF	—	Imperial Cancer Research Fund
IEEE	—	Institute of Electrical and Electronics Engineers
IG	—	IntelliGenetics
IGES	—	International Genetics Epidemiology Societies
IJCAI	—	International Joint Conference on Artificial Intelligence
IMA	—	Institute for Mathematics and its Applications
MBC	—	Massachusetts Biotechnology Council
MBL	—	Marine Biological Laboratory
MOD	—	March of Dimes
MOU	—	Memorandum of Understanding
MRC	—	Medical Research Council
NAPBC	—	Natl. Action Plan on Breast Cancer
NAS	—	Natl. Academy of Sciences
NASA	—	Natl. Aeronautics and Space Administration
NBAC	—	Natl. Bioethics Advisory Commission
NCGR	—	National Center for Genome Resources
NCI	—	Natl. Cancer Institute
NCSA	—	National Center for Supercomputing Applications

NCSL	—	National Conference of State Legislatures
NCSU	—	North Carolina State University
NEH	—	National Endowment for the Humanities
NFCR	—	National Foundation for Cancer Research
NFID	—	National Foundation for Infectious Diseases
NICHD	—	National Institute of Child Health and Human Development
NIGMS	—	National Institute of General Medical Sciences
NIH	—	National Institutes of Health
NIGMS	—	Natl. Institute of General Medical Sciences
NIST	—	Natl. Inst. of Standards and Technology
NLGLP	—	National Laboratory Gene Library Project
NLM	—	National Library of Medicine
NMHCC	—	National Managed Health Care Congress
NORD	—	National Organization for Rare Disorders
NRC	—	National Research Council
NRSA	—	National Research Service Award
NSF	—	Natl. Sci. Foundation
NSGC	—	National Society of Genetic Counselors

OHER	—	Office of Health and Environmental Research
OMIM	—	Online *Mendelian Inheritance in Man*
OPRR	—	Office of Protection from Res. Risks
ORAU	—	Oak Ridge Associated Universities
ORISE	—	Oak Ridge Inst. for Science and Education
ORF	—	open reading frame
ORNL	—	Oak Ridge National Laboratory
OSU	—	Oregon State University
PCR	—	Polymerase Chain Reaction
PFGE	—	Pulsed-field gel electrophresis
PG	—	Plant Genome
PIR	—	Protein Info. Resource
PNNL	—	Pacific Northwest National Laboratory
PRIM&R	—	Public Responsibility in Medicine & Research
RAPD	—	Random Amplified Polymorphic DNA
RARA	—	Retinoic Acid Receptor
RECOMB	—	Conference on Computational Molecular Biology
RFLP	—	Restriction Fragment Length Polymorphism
SCAN	—	Sequence Comparison ANalysis Program
SCE	—	School of Continuing Education

SCI	—	Society of Chemical Industry
SDC	—	San Diego Conference
SHOM	—	Sequencing by Hybridization on Matrices
SIMS	—	Societal Institute of the Mathematical Sciences
SIVB	—	Society for In Vitro Biology
SNP	—	Single Nucleotide Polymorphism
UICM	—	Univ. of Illinois College of Medicine

List of Glossary of Terms

Adaptive radiation: The evolution of new species or sub-species to fill unoccupied ecological niches.

Aerobe: A microorganism that grows in the presence of oxygen. See Anaerobe.

Agarose gel electrophoresis: A matrix composed of a highly purified form of agar that is used to separate larger DNA and RNA molecules ranging 20,000 nucleotides.

Agrobacterium: Agrobacteria is a soil bacteria that works as a natural genetic engineer. In nature, it forces plants to become good hosts by inserting genes into plant cells causing plants to make metabolites agrobacterium need for growth. Genetic engineers have taken advantage of agrobacteria's natural abilities by utilizing them to insert an altered piece of DNA into a host organism.

Biomotors: Driven by energy sources such as *adenosine triphosphate (ATP)* for chemical transduction and other processes. These biomotors are considered to be biomolecular and are discussed in the body of this report, but strictly speaking they do not conform to the panel's definition of self- assembly.

Biosensor: The term "biosensor" is a general designation that denotes either a sensor to detect a biological substance or a sensor which incorporates the use of biological molecules such as antibodies or enzymes. Biosensors are a subcategory of chemical sensors.

Clone: An exact genetic replica of a specific gene or an entire organism. See Cloning.

Cloning vector: A DNA molecule capable of autonomous replication within the cloning host cell (e.g *E. coli*). The vectors contain restriction enzyme sites for insertion of foreign DNA. Cloning vectors are derived from bacterial plasmids, bacteriophages, or viruses.

Competency: An ephemeral state, induced by treatment with cold cations, during which bacterial cells are capable of uptaking foreign DNA.

Concatemer: A DNA segment composed of repeated sequences linked end to end.

Conjugation: The joining of two bacteria cells when genetic material is transferred from one bacterium to another.

Coulomb [C]: Measure of electrical charge: 1 C is an amount of charge equal to that of about 6.24x1018 electrons.

Coupling: A connection between more than one object or energy pathway so that together they function as a single unit.

Crossing-over: The exchange of DNA sequences between chromatids of homologous chromosomes during meiosis.

Cross-pollination: Fertilization of a plant from a plant with a different genetic makeup.

Crosstalk: Electromagnetic noise transmitted between leads or circuits in close proximity to each other.

Enabling science and technologies: Areas of research relevant to a particular goal, such as nanotechnology. Also, technology that "enables" other technology to advance, such as the transistor enabled the computer chip revolution, as did photolithography.

Encapsidation: Process by which a virus' nucleic acid is enclosed in a capsid. See Coat protein.

Endocrine cell: Specialized animal cell that secretes a hormone into the blood; usually part of a gland, such as the thyroid or pituitary gland.

Endocytosis: Uptake of material into a cell by an invagination of the plasma membrane and its internalization in a membrane-bounded vesicle.

Endophyte: An organism that lives inside another.

Endoplasmic reticulum (ER): Labyrinthine, membrane-bounded compartment in the cytoplasm of eucaryotic cells, where lipids are synthesized and membrane-bound proteins are made.

Energy [J]: Capacity for performing work or to cause heat flow. Like work itself, it is measured in Joules.

Environmental Protection Agency (EPA): The U.S. regulatory agency for biotechnology of microbes. The major laws under which the agency has regulatory powers are the Federal Insecticide, Fungicide, and Rodenticide Act (FIFRA); and the Toxic Substances Control Act (TSCA).

Enzymes: Proteins that control the various steps in all chemical reactions.

Ethidium bromide: A fluorescent dye used to stain DNA and RNA. The dye fluoresces when exposed to UV light.

Eukaryote: An organism whose cells possess a nucleus and other membrane-bound vesicles, including all members of the protist, fungi, plant and animal kingdoms; and excluding viruses, bacteria, and blue-green algae. See Prokaryote.

Eurisko: A computer program developed by Professor Douglas Lenat which is able to apply heuristic rules for performing various tasks, including the invention of new heuristic rules.

FIFRA: The Federal Insecticide, Fungicide, and Rodenticide Act. See Environmental Protection Agency.

Fingerprint: A set of molecular markers sufficiently diverse to identify particular individuals with reasonable certainty.

Flanking region: The DNA sequences extending on either side of a specific locus or gene.

Food and Drug Administration (FDA): The U.S. agency responsible for regulation of biotechnology food products. The major laws under which the agency has regulatory powers include the Food, Drug, and Cosmetic Act; and the Public Health Service Act.

Gel, agarose: A substance used to separate DNA or RNA fragments by size. Used for southern and northern blots.

Gel, polyacrylamide: A substance used to separate proteins by size. Used for western blots.

GEM: A genetically engineered microorganism.

Gene amplification: The presence of multiple genes. Amplification is one mechanism through which proto-oncogenes are activated in malignant cells.

Gene cloning: The process of synthesizing multiple copies of a particular DNA sequence using a bacteria cell or another organism as a host. See DNA, Host.

Gene expression: The process of producing a protein from its DNA- and mRNA-coding sequences.

Gene flow: The exchange of genes between different but (usually) related populations.

Gene frequency: The percentage of a given allele in a population of organisms.

Gene gun: Transformation technique that uses accelerated particles coated with DNA to introduce foreign DNA into recipient plant.

Gene insertion: The addition of one or more copies of a normal gene into a defective chromosome.

Gene linkage: The hereditary association of genes located on the same chromosome.

Gene pool: The totality of all alleles of all genes of all individuals in a particular population.

Genetic drift: Random variation in gene frequency from one generation to another.

Genetic engineering: Genetic engineering is the process of manually adding new DNA on a molecular level with the goal of adding one or more new traits that are not already found in that organism. For the purposes of this web site, genetic engineering includes recombinant DNA and gene splicing technologies.

Genetic marker: A gene or group of genes used to "mark" or track the action of microbes.

Genome: The genetic complement contained in the chromosomes of a given organism, usually the haploid chromosome state. The full chromosome set containing all the genes of a particular individual.

Genomics:: The study of the structure and function of genomes. Genomics usually involves high speed sequencing of the DNA and computer searches for sequences that code for genes.

Genotype: The sum total of the genetic information of an organism including the linkage relationships between genes. The genotype, modified by environment, determines the phenotype.

Genus: A category including closely related species. Interbreeding between organisms within the same category can occur.

Germ cell (germ line) gene therapy: The repair or replacement of a defective gene within the gamete-forming tissues, which produces a heritable change in an organism's genetic constitution.

Green revolution: Advances in genetics, petrochemicals, and machinery that culminated in a dramatic increase in crop productivity during the third quarter of the 20th century.

GUI: Graphical User Interface — hardware, software, and firmware that produces the display on modern personal computers.

Heterochromatin: Dark-stained regions of chromosomes thought to be for the most part genetically inactive.

Heteroduplex: A double-stranded DNA molecule or DNA-RNA hybrid, where each strand is of a different origin.

Homologous chromosomes: Chromosomes that have the same linear arrangement of genes—a pair of matching chromosomes in a diploid organism. See Chromosomes.

Homologous recombination: The exchange of DNA fragments between two DNA molecules or chromatids of paired chromosomes (during crossing over) at the site of identical nucleotide sequences.

Homozygote: An organism whose genotype is characterized by two identical alleles of a gene. See Allele, Genotype.

Insulin: A peptide hormone secreted by the islets of Langerhans of the pancreas that regulates the level of sugar in the blood.

Interferon: A family of small proteins that stimulate viral resistance in cells.

Intergenic regions: DNA sequences located between genes that comprise a large percentage of the human genome with no known function.

Introgression: Backcrossing of hybrids of two plant populations to introduce new genes into a wild population.

Intron: A noncoding DNA sequence within a gene that is initially transcribed into messenger RNA but is later snipped out. See Coding, DNA, Messenger RNA, Transcription.

Kilohertz (kHz): One thousand cycles per second.

Kilojoule: Standard unit of energy equal to 1000 joules, or 0.24 kilocalories.

Lag phase: The initial growth phase, during which cell number remains relatively constant prior to rapid growth. See growth phase.

Legume: A member of the pea family that possesses root nodules containing nitrogen-fixing bacteria.

Library, DNA: A large set of clones of DNA fragments from a particular organism. DNA libraries are usually maintained in *E. coli* or in bacteriophages.

Ligation: Joining two fragments of DNA end to end. See sticky ends.

Linked genes/markers: Genes and/or markers that are so closely associated on the chromosome that they are coinherited in 80% or more of cases.

Linker: A short, double-stranded oligonucleotide containing a restriction endonuclease recognition site, which is ligated to the ends of a DNA fragment.

Locus (plural = loci): A specific location or site on a chromosome.

Logarithmic phase (log or exponential growth phase):

The steepest slope of the growth curve—the phase of vigorous growth during which cell number doubles every 20-30 minutes.

Lysogen: A bacterial cell whose chromosome contains integrated viral DNA.

Lytic: A phase of the virus life cycle during which the virus replicates within the host cell, releasing a new generation of viruses when the infected cell lyses.

Machine-phase chemistry: The chemistry of systems in which all potentially reactive moieties follow controlled trajectories (e.g., guided by molecular machines working in vacuum).

Macroscale: Larger than nanoscale; often implies a design that humans can directly interact with; too large to be built by a single assembler (one cubic micron of diamond contains 176 billion atoms).

Magnetostrictive material: A material that changes dimension in the presence of a magnetic field or generates a magnetic field when mechanically deformed.

Malignant: Having the properties of cancerous growth.

Manufacturing: The ability to make products, in this case ranging from clothing, to electronics, to medical devices, to books, to building materials, and much more.

Meat Machine: AKA Cabinet Beast. A box containing assemblers and raw material, within which is formed meat [or whatever else it was programmed to make].

Megabyte (MB): 220 (= 1,048,576, or about one million) bytes of information.

Megahertz (Mhz): One million cycles per second

Meiosis: The reduction division process by which haploid gametes and spores are formed, consisting of a single duplication of the genetic material followed by two mitotic divisions.

Metabolism: The biochemical processes that sustain a living cell or organism.

Metallothionein: A protective protein that binds heavy metals, such as cadmium and lead.

Micelles: Micelles are small, spherical structures composed of molecules that attract one another to reduce surface tension. The head of the molecule is hydrophilic, meaning it likes water, while the interior portion is hydrophobic, meaning it avoids water.

Microbial mats (biofilms): Layered groups or communities of microbial populations.

Microelectronics: Narrower terms: MEMS, nanoelectronics, optoelectronics, SED Single Electron Devices. Related terms: molecular electronics, semiconductors

Microencapsulation: Individually encapsulated small particles.

Microinjection: A means to introduce a solution of DNA, protein, or other soluble material into a cell using a fine microcapillary pipet.

Microsystem: A microscale machine that can sense information from the environment and act accordingly. Outside the U.S., it can also refer to microelectromechanical systems (MEMS).

Microtubule: A hollow cylindrical tube used for the transport of materials inside cells.

Miller indices: A set of 3 integers (4 for hexagonal) that designate crystallographic planes, as determined from reciprocals of fractional axial intercepts.

Millimeter: One thousandth of a meter, or about 1/26 of an inch.

Mitosis: The replication of a cell to form two daughter cells with identical sets of chromosomes.

Molecular - Nanotechnology: The technology of precisely-constructed molecular-scale machines; from nanometer: a billionth of a meter. For a good introduction, see What is Nanotechnology? or Ralph C. Merkle's nanotechnology page.

Molecular Beam Epitaxy: [MBE] Process used to make compound (multi-layer) semiconductors. Consists of depositing alternating layers of materials, layer by layer, one type after another (such as the semiconductors gallium arsenide and aluminum gallium arsenide).

Molecular biology: The study of the biochemical and molecular interactions within living cells.

Molecular genetics: The study of the flow and regulation of genetic information between DNA, RNA, and protein molecules.

Molecular markers: Genetic traits which are detectable on gels or blots and can be used to construct genetic maps. Molecular markers usually have no known function. RFLPs are molecular markers.

Molecular Wire: A molecular wire—the simplest electronic component—is a quasi-one-dimensional molecule that can transport charge carriers (electrons or holes) between its ends. [Michael D Ward]

Monoclonal antibodies: Immunoglobulin molecules of single- epitope specificity that are secreted by a clone of B cells.

Monoculture: The agricultural practice of cultivating crops consisting of genetically similar organisms.

Monogenic: Controlled by or associated with a single gene.

Mycorrhizae: Fungi that form symbiotic relationships with roots of more developed plants.

Nanoscience: Nanoscience is the science of nanoscale phenomena. This is a very complicated area, since Quantum Mechanics is often needed to describe the physical properties of atoms and molecules properly.

Nitrocellulose: A membrane used to immobilize DNA, RNA, or protein, which can then be probed with a labeled sequence or antibody.

Nitrogen fixation: The conversion of atmospheric nitrogen to biologically usable nitrates.

Occupational Safety and Health Administration (OSHA): One of the U.S. agencies responsible for regulation of biotechnology. The major law under which the agency has regulatory powers is the Occupational Safety and Health Act.

Oncogenesis: The progression of cytological, genetic, and cellular changes that culminate in a malignant tumor.

Open pollination: Pollination by wind, insects, or other natural mechanisms.

Operator: A prokaryotic regulatory element that interacts with a repressor to control the transcription of adjacent structural genes.

Organelle: A cell structure that carries out a specialized function in the life of a cell.

Origin of replication: The nucleotide sequence at which DNA synthesis is initiated.

Paleontology: The study of the fossil record of past geological periods and of the phylogenetic relationships between ancient and contemporary plant and animal species.

Parasitism: The closee association of two or more dissimilar organisms where the association is harmful to at least one. See Commensalism, Parasitism, Symbiosis.

Particle bombardment: Using metal particles to blast DNA into cells for the purpose of genetic engineering. A gene gun is used to propel the metal particles.

Pheromone: A hormone-like substance that is secreted into the environment.

Phosphatase: An enzyme that hydrolyzes esters of phosphoric acid, removing a phosphate group.

Phosphodiester bond: A bond in which a phosphate group joins adjacent carbons through ester linkages. A condensation reaction between adjacent nucleotides results in a phosphodiester bond between 3' and 5' carbons in DNA and RNA.

Phosphorylation: The addition of a phosphate group to a compound.

Piezoelectric material: A ferroelectric material in which an electrical potential difference is created due to mechanical deformation, or conversely, in which the application of a voltage causes dimensional changes in the material.

Population: A local group of organisms belonging to the same species and capable of interbreeding.

Primary cell: A cell or cell line taken directly from a living organism, which is not immortalized.

Primer: A short DNA or RNA fragment annealed to single-stranded DNA, from which DNA polymerase extends a new DNA strand to produce a duplex molecule.

Prokaryote: A bacterial cell lacking a true nucleus; its DNA is usually in one long strand. See Eukaryote.

Pronucleus: Either of the two haploid gamete nuclei just prior to their fusion in the fertilized ovum.

Protease: An enzyme that cleaves peptide bcnds that link amino acids in protein molecules.

Protein kinase: An enzyme that adds phosphate groups to a protein molecule at serine, threonine, or tyrosine residues.

Recessive gene: Characterized as having a phenotype expressed only when both copies of the gene are mutated or missing.

Recessive: Refers to the member of a pair of alleles that fails to be expressed in the phenotype of the organism when the dominant member is present. Also refers to the phenotype of an individual that has only the recessive allele.

Rhizobia: Bacteria in a symbiotic relationship with leguminous plants that results in nitrogen fixation. See Nitrogen fixation.

Rhizosphere: The soils region on and around plant roots.

Ribosome: Cellular organelle that is the site of protein synthesis during translation.

Ribosome-binding site: The region of an mRNA molecule that binds the ribosome to initiate translation.

Risk assessment: Defined as a formalized basis for the objective evaluation of risk in a manner in which assumptions and uncertainties are clearly considered and presented.

RNA: Ribonucleic acid. The three basic types are messenger RNA (mRNA), ribosomal RNA (rRNA), and transfer RNA (tRNA).

Roundup Ready: A trademark for plants that are genetically engineered to be resistant to the herbicide Roundup (technical name: glyphosate).

Salmonella: A genus of rod-shaped, gram-negative bacteria that are a common cause of food poisoning.

Satellite RNA (viroids): A small, self-splicing RNA molecule that accompanies several plant viruses, including tobacco ringspot virus.

Self-pollination: Pollen of one plant is transferred to the female part of the same plant or another plant with the same genetic makeup.

Semiconservative replication: During DNA duplication, each strand of a parent DNA molecule is a template for the synthesis of its new complementary strand. Thus, one half of a preexisting DNA molecule is conserved during each round of replication.

Sequence hypothesis: Francis Crick's seminal concept that genetic information exists as a linear DNA code; DNA and protein sequence are colinear.

Sequence: Noun: the particular order of nucleotides in a DNA or RNA fragment. Verb: to determine the particular order of nucleotides in a strand of DNA.

Sequence-tagged site (STS): A unique (single-copy) DNA sequence used as a mapping landmark on a chromosome.

Sexual reproduction: The process where two cells (gametes) fuse to form one hybrid, fertilized cell. See Asexual reproduction, Gamete, Hybrid.

Signal transduction: The biochemical events that conduct the signal of a hormone or growth factor from the cell exterior, through the cell membrane, and into the cytoplasm. This involves a number of molecules, including receptors, proteins, and messengers.

Site-directed mutagenesis: The process of introducing spe- cific base-pair mutations into a gene.

Species: A classification of related organisms that can freely interbreed.

Spore: A form taken by certain microbes that enables them to exist in a dormant stage. It is an asexual reproductive cell.

Supergene: A group of neighboring genes on a chromosome that tend to be inherited together and sometimes are functionally related.

Supernatant: The soluble liquid &action of a sample after centrifugation or precipitation of insoluble solids.

Susceptibility: When individual or species will become ill or possibly die in the face of the stressful event or circumstances.

Transgenic plant: Genetically engineered plant or offspring of genetically engineered plants. The transgenic plant usually contains material from at least one unrelated organisms, such as from a virus, animal, or other plant. See Transgenic.

Upstream: The region extending in a 5' direction from a gene.

Vaccinia: The cowpox virus used to vaccinate against smallpox and, experimentally, as a carrier of genes for antigenic determinants cloned from other disease organisms.

Vasculography: A physical description (and mapping) of the human vascular system.

1

Biotechnology Ethics: Issues and Views

Ethical Considerations

- We could choose to have changes made to us, but we might also be making the choice for our children if the changes are carried through to the germline. Do we have that right, and how far should we take our ability?
- Conversely, is it responsible and ethically acceptable to leave the potentials of our children to the chance effects of the "genetic lottery", if we obtain the technological capacity to make positive changes?
- If genetic engineering became the way of the future, would people whose parents could not afford to genetically 'modify' them while still in an embryo, have a chance of achieving with high standards compared to the people who were 'modified' to be perfect?
- Is it ethical to experiment on embryos that have yet to be born?
- How would genetic engineering be used to revolutionize warfare?
- Who decides which changes will be made?

Crossing Species Boundaries

First, from a biological perspective, it is surprisingly difficult to answer the question What does it mean to cross species boundaries? This is true not only because of the number of species concepts (according to some, as many as 22), but also because species boundaries are not fixed.

- Species concepts: One classic definition of species is the biological species concept. This definition emphasizes the importance of reproductive isolation or lack of genetic exchange that separates species. By this account, crossing species boundaries would involve the transfer of genetic materials between populations of organisms that do not interbreed. In cases where such interbreeding can be achieved artificially, as in the laboratory, the raison d'être of the biological species concept is undermined. Other accounts of species may be brought to bear in place of the biological species concept, but the consensus among biologists is that no single species concept will be sufficient for all situations.
- Species boundaries: One of the consequences of our evolutionary past is that genes, gene regulatory networks, epigenetic developmental processes, and features of the biophysical environment are widely shared by different kinds of creatures. The idea of fixed or rigid breaks between species plays no role whatsoever in contemporary biology. Indeed, the fluidity of species boundaries has been revealed through the techniques of comparative genomics, warning against the interpretation of species as unique types.

Given the difficulty in defining species once and for all, and also the flexibility of species boundaries, what does it mean to cross species boundaries? When we refer to species and the crossing of species boundaries, we do so based on the following simple idea: Every individual human contains a human genome. In all likelihood, this genome will not be representative of other human genomes and will contain a lot of DNA that is contained in many other kinds of organisms, thanks to our evolution from a common ancestor. The same will be true with nonhuman organisms, such as a rose or a rat or a *Rhesus macaque*. As such, when we refer to crossing species boundaries, we refer to the combination of genetic or cellular material from two organisms that would generally be understood, in lay terms, as belonging to different species: a human as understood by most lay people, a *Rhesus macaque* as understood by most lay people, and so on.

Hybrids, Transgenics, and Chimeras

There are many types of interspecies organisms including hybrids, transgenics, and chimeras, each of which is created through different sorts of processes. Hybrids are created through breeding. Transgenics are produced through genetic manipulation and modification. Chimeras are the result of cell or tissue transplants.

- *Hybrids* are created by breeding across species. Hybrids are generally the result of combining an egg from one species with sperm from another to form a single embryo. Hybrids contain recombined genetic material throughout their genome and throughout all the tissues in their body.
- *Transgenics* are the result of gene transfer. Typically, transgenics contain transferred or manipulated genes in addition to the host nuclear and mitochondrial DNA. One exception may be a transgenic embryo comprised of the entire complement of nuclear DNA from one organism fused with an enucleated egg cell from another.
- *Chimeras* comprise a mixture of cells from two or more genetically distinct organisms of the same or different species. They are mosaics at the cellular level; individual cells are derived from either the host or the donor but not both.

Note that chimeras and transgenics need not cross species boundaries, whereas hybrids are always interspecific.

Multiple Applications

Crossing species boundaries happens all the time in nature and in agricultural settings, and it has a long history in developmental biology and immunology laboratories. Consider just a few examples:

- lateral gene transfer between bacteria, whereby genetic material is transmitted horizontally from one organism to another
- the crossing of strains of wheat; the insertion of genes from a plant (or an animal) into a plant to improve crop yield and robustness
- the mating of a horse with a donkey to create a mule; the fusion of sheep and goat cells to create a "geep"
- the transplantation of cells and tissues from one species of frog to another, and of neural tissue from quails to chickens, to study the complex processes of development

Crossing species boundaries between human and nonhuman animals is also commonplace:

- Animal tissues, cells, and their derivatives are often transferred to humans, whether by using insulin produced from pig or cow pancreases, injecting flu vaccines cultured in fertilized chicken eggs, or transplanting heart valves from pigs into humans.
- Human genes and cells are often transferred to animal hosts to create humanized animal models (such as OncoMouse, which develops human cancers), to grow humanized tissues that can be transplanted back into humans (such as sheep with human livers), or to test the developmental potential of transplants.

Stem Cell Research

In the first few years of human stem cell research, most of the ethics discussions centered on the use of human embryos as sources of stem cells. Research to derive human embryonic stem (hES) cells involves removing cells from the inner cell mass of a human blastocyst, which destroys the developing embryo. While this debate continued, further debate on the ethics of cloning to produce children, and cloning for biomedical research, emerged. This debate arose in response to claims about the anticipated benefit of future stem cell therapies using cells from cloned embryos, which would allow patients to receive transplants of cells containing their own DNA. While these issues remain ethically contentious, a new debate has recently emerged concerning the ethics of crossing species boundaries to make part-human chimeras. Stem cell scientists and others insist that this cross-species work is important to basic science and a necessary step on the path to regenerative medicine. They maintain that it would be unethical to involve humans in stem cell transplantation research without first having studied the safety and efficacy of the human cells in nonhuman animals.

Ethical Controversy

Issues of health and safety, especially given the possibility of zoonosis, or the transfer of a disease from nonhuman animals to humans, have long been front and center in the ethics debate about cross-species work. In the last decade or so, with the increase in science options for the crossing of species boundaries, other ethical issues have come to the fore. In 1997, Stuart Newman, a developmental biologist sponsored by biotechnology activist Jeremy Rifkin, sought to preclude the creation of a humanzee—a part-human, part-chimpanzee chimera. Together, Newman and Rifkin tried to patent the relevant technology so that they would be able to restrict its use and to promote a vigorous social dialogue about the desirability of such part-human beings. They were unsuccessful in obtaining the patent, leaving open the possibility that humanzees may soon walk among us, with or without patent protection. Examples of recent research involving the transplantation of cells and tissues into prenatal nonhuman animals (embryos and fetuses), the transplantation of cells and tissues into nonhuman animal brains, and the transplantation of cells and tissues into the brains of nonhuman primates, serve to make this point:

- Scientists at Harvard University have published their research involving the transfer of human neural stem cells into the developing fetal brain of bonnet monkeys.
- Scientists in Israel have reported that human embryonic stem cells transplanted into chick embryos differentiated into neurons.
- Scientists in Nevada have reported on inserting human neural stem cells into fetal sheep to assess their developmental potential.
- Scientists in California have reported on the development of functional neurons in mouse brains, where the neurons were derived from human embryonic stem cells.

While this research is ongoing, a debate has erupted about the ethics of creating part-human beings in response to proposals from some stem cell scientists to use nonhuman primates as an assay system for testing the developmental potential of human stem cells. The fact that biologists are especially interested in transplanting human neural stem cells into the brains of nonhuman primates intensifies the controversy about humanzee-like chimeras.

The Ethics of Creating Part-Human Beings

The ethical debate has been multivocal, with moral considerations raised from many perspectives, both religious and secular. The central moral concerns with creating part-human beings include worries about the following:

- the unnaturalness and intuitive repugnance of certain kinds of creatures, such as part-human combinations
- the threat of intensified moral confusion regarding the creation of novel part-human beings who violate the pragmatically clear moral demarcation line between species upon which current institutions, structures, and social practices are based
- the potential for transferring moral status to nonhuman animals by conferring on them characteristically human cognitive capacities, which may or may not threaten human dignity
- the possibility that enhanced animals would deserve to be treated as if they were human subjects but would continue being treated as if they were unenhanced nonhuman animals
- the moral status of nonhuman animals, especially primates, as experimental animals.

Scientists and others who advocate cross-species work argue that the part-human animals will be useful as disease models, assay systems, or organ sources. They dismiss the worries about repugnance, deny the potential for moral confusion, and endeavor to sidestep concerns about moral status and potential threats to human dignity. They also rely heavily on current norms for research involving humans to legitimate preclinical cross-species research in nonhuman animals. Additionally, advocates of *transhumanism*, the movement to enhance humans using biotechnology, argue that the creation of hybrids, transgenics, and chimeras may be useful in the quest to radically alter humans. They see an opportunity to improve upon human nature and to enhance cognitive and physical performance—an idea that is itself morally controversial.

Toward a Constructive Public Debate

Unfortunately, much of the public debate on the ethics of crossing species boundaries is characterized by sensationalism and political posturing. While some commentators have attempted to explore the moral dimensions of interspecies research in careful and respectful terms, many of the media reports have exaggerated the conflict, providing more heat than light. Even so, attempts at public education and public engagement have tended to reveal the persistence of the moral controversy. This suggests the need for scientists, ethicists, and others to take seriously the ethical concerns that have been raised. The voluntary guidelines for human embryonic stem cell research recently published by the National Academy of Sciences arguably are an attempt to do just this. As we have argued elsewhere, however, considerably more debate and discussion is needed about the fundamental underlying values.

ETHICAL ASPECTS

- Is biotechnology morally good or bad?
- Is working to create biotechnology morally right or wrong?

- It depends.

Some Practical Ethical Theories

- Reasonable Person Utilitarianism (Do the best you can.):

An act is morally right only if a reasonable person would reasonably believe that the utility of the consequences of the act will probably be as great as any alternative to the action at that time.

- Quasi-Categorical Imperative (If you do not want it done to you, then do not do it to anyone else AND if it is wrong for everyone to do it, then it is wrong for you to do it):

An act is morally right only if in doing the act, the agent does not treat anyone as a mere means.

The Moral Saint Test

- Think of an actual person who you who you would like to emulate.
- Eliminate the person's vices.
- In your mind, tell the person what you think is the correct solution to moral dilemma. The correct solution must be one that you would do if you were faced with the choice.
- If your paradigm would not lose some respect for you, then you know that you probably have selected the correct solution.
- If your paradigm would lose some respect for you, then you know that you probably have selected the wrong solution.

Applying the Theory

- Who is affected by the action?
- How are those people affected?
- Which alternative is likely to be the best?
- Are the mental states, including but not limited to intentions and motives, of the agent(s) performing the actions primarily good?

Applying the Theory

- Golden Rice
  - o Unethical.
  - o RPU – To a reasonable person, the outcome did not reasonably appear to be the best that could have been done.

- o QCI – The desires of the individuals who were to benefit were not consulted.

Applying the Theory

- Pharmaceutical companies now use genetic engineering to produce large quantities of insulin for diabetics around the world. The common Escherichia coli (E. coli) bacteria is used to produce insulin. Using recombinant DNA (glossary id 27) technology, the gene for producing insulin is inserted into the E. coli's genetic material. These genetically engineered bacteria are turned into tiny insulin producing factories. http://www3.iptv.org/exploremore/ge/uses/use2_medical.cfm#

Applying the Theory

- Artificial Insulin
  - — Ethical
  - — RPU – Many people are benefited with none harmed.
  - — QCI – No one is treated as a mere means.

Social & Ethical Issues in Nano Biotechnology

At some point in the near future, nanomachines will produce consumer goods, rather than the bulky machines that are used today. NT works "…by piecing together one atom or molecule at a time…" For nanotechnology (NT) to work, small devices called assemblers must be created in order to combine atoms that will produce consumer goods. Replicators will also have to be made to produce enough assemblers that will be required to create goods. Of the many positive features of NT, nanomachines will produce: stronger products than those produced by traditional methods; food and water will be able to be replicated, thus, eliminating hunger worldwide; computers could become dramatically more powerful without any increase in size; nanosurgeons could undertake delicate operations much more precise than current surgery is able to; and nanorobots could be used to alter one's physical features such as ears, nose, eye colour, etc. Nanotechnology could also have many positive effects on the environment, including: reconstruction of the ozone layer; quickly cleaning up oil spills; purifying water; and eliminating the use of non-renewable resources (which could be made by nanomachines). Brown, Doug *"Nano Litterbugs? Experts See Potential Pollution Problems."* (From www.smalltimes.com). The U.S. Environmental Protection Agency (EPA) has begun research into the possible negative impacts associated with NT. Even as some products using NT have begun to be widely distributed, the environmental impact is still unknown. A concern for Vicki Colvin, who is director of The Center for Biological and Environmental Nanotechnology, is that no research is being conducted into how to control nanomaterial if it was discovered to be harmful to humans or the environment. Based on preliminary research on animals, it is known that nanomaterials do accumulate within the body. Also, nanomaterials can be overtaken by bacteria, which could mean nanomaterials could infiltrate the food chain. Bacteria enhanced by nanomaterials could produce more potent bacteria. While NT promises to have numerous benefits for society, many of the potential risks remain uninvestigated.

Joy, Bill *"Why the future doesn't need us." (*In *Wired* magazine. *April, 2000).* Joy's concern over NT is amplified, due to the fact that engineered organisms can self-replicate. Such replication could lead to situations where things could quickly get out of hand. "A bomb is blown up only once—but one bot can become many, and quickly get out of control." An additional worry for Joy is, unlike weapons of mass destruction (WMD) – which require hard-to-get materials and highly specialized knowledge – genetics, nanotechnology, and robotics (GNR) are available to a wide range of people, thus making them all the more dangerous. While WMD are strictly controlled by the military, GNR's are corporate sponsored. Such a wide dispersion of NT increases the risk. Joy emphasizes prudence in our development of NT, because if we proceed too quickly, we could cause our own extinction. After all, we are still unable to control the spread of viruses and other undesirable components of life. Joy draws a parallel between NT and the atomic bomb. The creation of the atomic bomb led to the development of an arms race that lasted nearly half a century and had many detrimental effects. Atomic and nuclear weapons have the ability to destroy the world many times over, as does NT. We should not repeat the same mistakes with NT that we did with WMD. "The only realistic alternative I see is relinquishment: to limit development of the technologies that are too dangerous, by limiting our pursuit of certain kinds of knowledge." Limiting access to critical information involving NT must be enforced because it contains such a destructive force to humanity (e.g. exterminating the entire race). People will inevitably object to the infringement on their right to privacy and pursuit of intellectual property, but this is a necessary step if we are to avoid the dangers NT poses. Additionally, scientists will have to comply with strict ethical codes to ensure compliance with relinquishment rules. While humanity's dream of immortality will likely continue forever, certain limits to the means creating this end must be imposed due to the inherent danger.

Mehta, Michael D. "On Nano-Panopticism: A Sociological Perspective." This article deals with privacy issues associated with surveillance, and how surveillance will be changed with the development of NT. It has been scientifically documented that observation of a subject will change the regular behaviour of that subject. For example, prisoners who are held in prison cells that are in open view of a prison guard behave better than those prisoners who are afforded more privacy. Therefore, if this concept is applied to society as a whole, one can see the ramifications increased surveillance might have on public behaviour. Increased surveillance with the use of NT may have many damaging effects that may express themselves through a decrease in public trust, less involvement in public life, and may result ultimately in an Orwellian state. Mehta offers several suggestions on how to deal with the dangers related to nano-panopticism. They are: develop an independent agency to monitor and regulate progress in NT; introduce tough laws to ensure that privacy is protected; a review of a corporation's code of conducts to ensure nano-panopticism does not become a problem; public consultation of the appropriate uses of NT; and finally, increase the awareness of scientists so that they are better able to understand the impact new technologies have on society.

Mnyusiwalla, A. Abdallah S. Daar, & Peter A. Singer "'Mind the gap': science and ethics in nanotechnology." (From Nanotechnology, February 13, 2003). As NT quickly develops, the ethical evaluation of such a development has yet to begin. Research into NT's impact on ethical, environmental, economic, legal, and social implications (NE^3LS) must try to keep pace with the technological progress that has been made. Otherwise, the technological progress will slow down. K. Eric Drexler, author of *Engines of Creation*, and currently working for the Foresight Institute, has begun to develop guidelines for the use of NT, particularly concerning self-replication, but also issues such as wealth distribution and environmental protection. While there is a call for increased research into the effects of NT on society, the environment, etc, and furthermore, that the funds for such research are available, very little research has been done. For example, The National Nanotechnology Initiative (NNI) spent less than half of its budget on studying the impacts NT will have on society. In addition, the National Science Foundation (NSF) did not fund any research into the societal implications of NT due to lack of research grant proposals. The ethical issues associated with NT fall into a variety of categories including:

1. *Equity*: NT does not stand to help developed countries only, but also (and perhaps mainly) developing countries. For example, third world countries suffer most from

things that can be improved upon from advancements made in NT. For instance, providing cleaner water, developing cheaper energy, and also the enormous health benefits to be reaped from NT are all aspects of NT that could have a dramatic impact on third world countries.

2. *Privacy and security*: While NT could dramatically improve surveillance systems and would undoubtedly have numerous military applications, some are left wondering how personal privacy would be affected. Questions concerning the regulation of this new technology are arising without many answers being offered.
3. *Environment*: The effects of NT materials on the environment are unknown. The EPA has increased funding to study the effects of NT on the environment.
4. *Human or machine?* How far are humans willing to go with replacement of human parts with replacement robot parts?

Earlier lessons from genomics and biotechnology (recommendations):

- A percentage of the research and development funds should go to study the societal impacts of NT. In the *Human Genome Project*, 3-5% of the total budget was dedicated to studying the societal impacts.
- Large-scale interdisciplinary research platforms should be avoided.
- The capacity for research should be increased. This would include research grants, career awards, increased funding, etc.
- There should be a cross-section of groups working on the implications of NT in order to have a variety of views expressed.
- A global opinion-leaders network for social and ethical

implications ought to be established so that third world countries may be involved.

- Public engagement is crucial if NT is to have the impact it potentially holds. The public needs to have accurate information concerning the social and ethical implications of NT. Science museums could include exhibits on NT; as well, schools could discuss the issue in depth.

Moor, James & John Weckert *"Nanoethics: Assessing the Nanoscale from an Ethical Point of View."* Weckert and Moor argue that nanoethics cannot develop fully or effectively until the technology develops further and its consequences become known. Weckert and Moor do acknowledge the potentially enormous impact NT will have on society, but are hesitant to develop strategies to deal with problems that do not yet exist. They recommend continued monitoring of NT. Pagán, Sylvia *"DNA nanoballs boost gene therapy."* (In *New Scientist*, May, 2002). NT is now beginning to enhance older technologies used to treat patients with cystic fibrosis. Nanoparticles made up of a single DNA molecule are much smaller, and therefore much better able to reach the nucleus of a cell, than were previous methods involving modified viruses. NT shows promising signs of contributing to the advancement of the cure for cystic fibrosis. Phoenix, Chris & Mike Treder *"Applying the Precautionary Principle to Nanotechnology."* (From *The Center for Responsible Nanotechnology*, January, 2003). There are two forms of the precautionary principle: (1) the strict form; and (2) the active form. *Strict form*: "...requires inaction when action might pose a risk." *Active form*: "...calls for choosing less risky alternatives when they are available, and for taking responsibility for potential risks." One shortcoming of the strict form of the Precautionary Principle is that most actions carry some sort of risk. If we follow the strict form, much of the scientific research currently underway would have to be halted. Second, sometimes inaction can be more harmful than action. Sometimes the benefits to be gained by scientific research outweigh the minimal risks involved. The Center for Responsible Nanotechnology (CRN) believes that a total ban on development of NT would not be practical; enforcement of such a ban would be nearly impossible. Not only is such a ban not practical, it may not be desired. There are many benefits that could be brought about by NT, including increased health, less environmental damage, cheap and efficient production of goods, etc. These factors must be considered when determining the fate of NT. The CRN believes that the strict form of the Precautionary Principle is inadequate for dealing with NT. This is so because inaction carries with it certain risks:

1. No other solution may be found for certain pressing problems.
2. Inaction on the part of responsible people could simply lead to the development and use of MNT by less responsible people.
3. Lack of understanding of the technology will leave the world ill-equipped to deal with irresponsible use.

The CRN does, however, believe that the active form of the Precautionary. Principle is adequate in dealing with research into NT. Because the active form seeks to develop the least risky solution to existing problems, this is more desirable than the strict form. The final recommendation of the CRN is to create "...one–*and only one*–molecular nanotechnology program, [which would permit] the widespread but restricted use of the resulting manufacturing capability. *Phoenix, Chris & Mike Treder* "Safe Utilization of Advanced Nanotechnology." (From The Center for Responsible Nanotechnology, January, 2003). There are many risks associated with NT. For instance, Drexler's "gray goo" problem–where self-replicating nanoassemblers multiply out of control; NT falling into the hands of rogue states; and decreased privacy due to the sizing-down of surveillance equipment, to name a few. Given these risks, it would appear that some regulation of NT is in order. If NT is regulated too strictly this could lead to the development of a black market where irresponsible users could do a great deal of damage. Phoenix and Treder propose a two-dimensional approach to the restriction of NT which will reduce risks associated with NT, yet allow the technology to further develop. First, products that use NT do not necessarily pose any extraordinary risks, they are simply products that have been enhanced in some way by the use of NT; they will not be able to self-replicate whatsoever. However, the factories used to produce NT products could quite easily be very dangerous. NT factories, if controlled by the wrong people, could hold great danger for humankind with the potential to develop destructive weapons much greater than are available today. Phoenix and Treder stress that "...research leading to advanced nanotechnology will have to be carefully monitored and controlled. However, the same *is not true of product research and development*." If NT factories are to be safe, certain precautions must be taken. First, assemblers need not ever leave the factory, they can easily be fastened to the factory to avoid them escaping into the environment and getting out of control. Second, factories can be programmed to produce only specified products that are known to be safe. Third, one could go so far as to build in a tracking device to ensure that NT products were being used safely and for their intended purpose. Furthermore, nanofactories could be made tamper-proof so that they could shut down if attempt at alteration was made. Finally, a nanofactory could be connected to a central controller which would then determine if the desired product was safe. If the central controller determined the product to be unsafe, the controller could disable the factory. Additional advantages of this scenario would be that the controller could automatically provide updates to the nanofactories and also relay the location of the factory.

Rotman, David "Measuring the Risks of Nanotechnology." (In Technology Review, April, 2003).

Vicki Colvin, Director of The Center for Biological and Environmental Nanotechnology at Rice University, states that the possible dangers associated with NT are largely unknown at this point. Despite the fact that nanomaterials have been put to use in products such as sunscreen, the potential risks of these materials is unknown. It is believed that some nanoparticles can be transmitted from the blood stream to brain material. Colvin believes that the benefits of NT–especially its medical applications–far outweigh any risks that have been hypothesized. Groups such as Greenpeace have begun to investigate NT for any risks it may pose.

"Handle Nano-Technology with Care," (In African Business. March, 2003). [not available on-line]. The action group on Erosion, Technology and Concentration (ETC) is worried that developments in NT are going unnoticed both by the public and government regulatory agencies. This is of great concern given that NT will have such a profound impact on society. One problem with regulating NT is that it has such a wide variety of applications and is therefore difficult to control under one regulatory agency. Furthermore, because many of the products that have incorporated NT merely use the same compounds that were previously used but on a lower scale, many people do not recognize any associated risks whatsoever. ETC is calling for a worldwide evaluation of the potential impacts of NT, and until such a process has begun, production of nanomaterials ought to cease.

THE ETHICS OF BIOTECHNOLOGY: A PERSONAL REFLECTION OF VIEWS

...(This is my short opinion about this issue not a major essay and just as straight to the point essay as I may go)...

*What is ethical and what isn't?, as far as I am concerned there are much more unethical issues in society today than putting a ban on genetical engineering... And by the way, one of the meanings of health is the balance between mind, body and your animic state (spirit), which is a state of your mind...

So the ethical men and women want to put a ban on gene therapy unless it is used to treat a deadly illness... So let me ask you Mrs. Ethics, did you marry your husband because he was Medium built or athletic (in other words not fat because of some genetic disorder caused by their tyroid), or because he was Tall (and not a person with genes that determined him to be a man of short stature certainly not shorter than yourself Mrs.Ethics - much less a dwarf-.By the way Mrs. Ethics did you go to church last Sunday, ... And to Mr. Ethics, did you choose your wife because she was not fat because of some genetic disorder caused by their Tyroid gland; or because she had a beautiful voice attached to a nice ass. Oh! and may god forbid that Mr. Ethics purchases one of the penis enlargement pills so that he might be a better. So that he might have a better time with Mrs. beatiful voice and her nice rear attachment. And by the way Mr. Ethics did you go to church last Sunday so that you might showoff your public Ethical behavior with Mrs. Ethics...

...I hope that my point is getting accross. But for as long as any of you Mr. Ethics and Mrs. Ethics, have discriminated in you choice of mate for any of the above mentioned reasons then you Sir or Mrs. have NO MORAL VALUE TO BAN ANY GENETIC RESEARCH OF ANY KIND WHATSOEVER, and if you do oppose, and you do believe in GOD then fear for him because you people are total hypocrits... ... The reason why I am writing this short essay is because I have discriminated against someone, just like yourself (more than likely)and I am pretty sure that people have discriminated against me for one or another reason, but what if people could use genetic treatments for cosmetic issues, what's wrong with that? Why should a woman go for the pain of a breast implant, or a liposuction (which is not only painful but risky) or how about the painful limb lenghtening that some women and men are using just to be taller and have a better job or a date... When those problems could be solved by putting the right genes on the people who desire those treatments? Why should you morons put a ban on something like that!!! Are you people crazy!!! How for heaven's sake a gene pool of six billion is going to become ill prone when the pool of the 12 tribes of israel evolved to this six billion pool? I mean common people! Do you people have a brain inside your head or you head is just there to collect dandruff?... To you people it is ethical to allow our fellow human beings to go through the pain of limb lenghtening, liposuction, or breast implant but it is not ethical to find a gene therapy to treat the situation?...

To the Ethic group of people who want to put a ban on this ongoing research, I'd like to say that you are the most unethical people on this planet. And to you genetic scientists my heart goes out to you, and may God bless you for your remarkable job. And I hope that these Ethical people don't become a hurdle in your quest.

ETHICS OF STEM CELL RESEARCH : A CASE STUDY

The Moral Status of the Embryo

There is little doubt that public reflection on stem cell research in the United States has been affected by the extraordinarily volatile cross-currents of the abortion debate. Although I will indicate below several reasons why framing the stem cell debate as a subset of that on abortion is problematic, nevertheless, in its current form, stem cell research is debated in terms dictated by the abortion controversy, and that has meant that questions about the status of the embryo have been particularly prominent.1 For example, the National Bioethics Advisory Commission (NBAC) described the ethical issues raised by stem cell research as "principally related to the current sources and/or methods of deriving these cells" (NBAC, 1999, 45). A policy brief from the American Association for the Advancement of Science (AAAS) begins its discussion of the ethical dispute over stem cell research by citing the disagreement over the status of the embryo as the decisive variable leading to fundamentally different views on this research (AAAS, 1999, 11).

The National Academy of Sciences's report on stem cell research claims that "the most basic [ethical] objection to embryonic stem cell research is rooted in the fact that such research deprives a human embryo of any further potential to develop into a complete human being" (National Academy of Sciences, 2002, 44). The Ethics Advisory Board of the Geron Corporation lists the moral status question as the first moral consideration relevant to deciding the acceptability of stem cell research (Geron Corporation Ethics Advisory Board, 1999, 32). The list could go on.

Despite the fact that these statements all insist on the importance of the status question, they also recognize that the debate about the status of the early embryo is not new and that the controversy over stem cell research does not, strictly speaking, raise novel issues in this regard. Indeed, it is probably best to place the nitial skirmishes over stem cell research in the context of moral debates about human embryo research generally. In fact, it is worth noting that the report of the NIH Human Embryo Research Panel (HERP), published in 1994, explicitly identified the isolation of human embryonic stem cells as one of thirteen areas of research with preimplantation embryos that might yield significant scientific benefit and that should be considered for federal funding (See NIH HERP, 1994, ch. 2).

Although the recommendations of the HERP were never implemented, the fact that a high-profile panel reviewed ex utero preimplantation human embryo research and explicitly endorsed stem cell research, meant that the panel report would affect the policy debate about stem cell research, even though its recommendation that the derivation and use of stem cells be federally funded was not adopted. For one thing, the panel's anticipatory support for stem cell research assured that when human stem cells were actually derived several years later, the debate that ensued would be tied to the abortion controversy. As members of the HERP panel have made clear, from the start, the work of the panel was embroiled in controversy. For example, shortly after the HERP was impaneled, thirty-two members of Congress wrote to Harold Varmus, the director of NIH, to complain about the composition of the panel. A lawsuit was filed in an attempt to prevent the panel from meeting, and members of the panel received threatening letters and phone calls (Green, 1994; Tauer, 1995; Hall, 2003).

Given the pro-life opposition to the HERP panel and its recommendations, it is no real surprise that initial reactions to the prospect of human stem cell research fell out along the fault lines of abortion politics in the country. By and large, individuals and groups opposed to abortion tended to be opposed to stem cell research, and individuals and groups supportive of legalized abortion tended to support stem cell research.2 For example, the testimony that Richard Doerflinger, the principal spokesperson for the U.S. Catholic bishops on pro-life matters, offered before the Senate Appropriation Subcommittee on Labor, Health, and Education in 1998 was substantially the same as that he offered before the HERP in 1994 on stem cell research (Doerflinger, 1998, 1994). In both cases, the fundamental issue was the status of the embryo. Given Catholic teaching that the embryo must be treated as a person from conception, no experimentation on the embryo can be allowed that would not also be allowed on infants or children. Hence, the Catholic church treats stem cell research as it has treated previous issues involving the destruction of human embryos; it is condemned as morally abhorrent.

In similar fashion, the arguments reviewed by the HERP panel that supported embryo research generally in 1994, were mobilized again four years later when stem cell research was the specific point of contention; and again the focal point was embryo status.

Consequently, just as the HERP report opted for a "pluralistic" view of the embryo that emphasized its developmental potential, so, too, did the NBAC endorse the idea that the early embryo deserves respect, but is not to be treated fully as a person.3 Moreover, the fact that the HERP defended its support of stem cell research by stressing the developmental capacity of the embryo also shaped the trajectory of much subsequent support for this work, because insisting on respect for the embryo but denying its personhood meant explaining how one could respect the embryo while nevertheless destroying it. Daniel Callahan, for example, posed this problem very strongly in response to the HERP report. If "profound respect" for the embryo is compatible with destroying it, he asked, "What in the world can that kind of respect mean?" It is, he says, "an odd form of esteem, at once high-minded and altogether lethal" (Callahan, 1995). Callahan was not alone in raising this issue and attempts to answer his question continue to appear in the literature (See Lebacqz, 2001; Meyer and Nelson, 2001; Ryan, "Creating Embryos," 2001; Steinbock, 2001, 2000).

In retrospect, then, it seems that the HERP report served almost as choreography for the initial debates about stem cell research, and, as a result, the steps in the debate closely followed those that are familiar from the abortion controversy (On this point, see Hall 2003). The upshot, in my view, is that much of the debate has been too narrowly focused and has a kind of repetitive and rigid quality to it. As I noted above, for example, the Catholic church has repeatedly claimed that the central issue raised by stem cell research is that it involves the destruction of human embryos, embryos it believes should be treated as persons.4 For that reason, the rhetoric with which the Catholic church condemns embryonic stem cell research closely parallels that used to condemn abortion. Yet, because the American bishops do not want to be perceived as antiscience, they have also repeatedly and uncritically praised adult stem cell research, even though there are good reasons, given Catholic concerns about social justice, to be concerned about the pursuit of adult stem cell research. I will return to this point below, but for now I wish simply to note that much of the opposition to embryonic stem cell work has resembled Catholic opposition in being circumscribed by questions of embryo status, narrowly construed.

A similar constriction, however, is also apparent in the preoccupations of supporters of stem cell research. Just as pponents of this research have ritualistically condemned the destruction of early embryos but uncritically celebrated adult stem cell work, supporters of embryonic stem cell research have typically insisted on using embryos left over from IVF procedures, while repudiating the use of embryos created solely for research. Indeed, insisting on the distinction between so-called "spare" embryos and "research" embryos and endorsing only the use of spare embryos has been one way that supporters of embryo research have tried to demonstrate their "respect" for the embryo. Yet, it is worth asking whether the spare embryo/research embryo distinction does not, to borrow Daniel Callahan's image, provide a kind of "wafting incense" to mask what supporters still find a disquieting smell (Callahan, 1995).

Although the debate about stem cell research might have been framed in terms of the abortion controversy in any event, the HERP report insured that the initial debate over stem cell work that followed in aftermath of the public announcement of the work of John Gearhart (Shamblott et al., 1998) and James Thomson in 1998 (Thomson et al.) would be navigated in the wake of the conflict over abortion. As I indicated, the upshot is that the discussion about stem cell research has been more cramped than it might otherwise have been. The discussion has been too focused on the details of embryological development; too focused on the differences between those who view the early embryo as a person and those who do not; and far too individualistically oriented. Before turning to ways that the debate might be become less cramped, let me focus more concretely on these difficulties.

The point about the debate being framed too individualistically is nicely illustrated in an article on abortion by Lisa Sowle Cahill entitled "Abortion, Autonomy, and Community" (Cahill, 1996). Cahill begins this article by claiming that, in discussing the morality of abortion, there is no way to avoid the question of the status of the fetus. Nevertheless, she says, the debate about fetal status is almost always conducted with the goal of determining the rights involved, where rights are understood very individualistically. To the degree that the fetus is acknowledged to have rights, those rights are pitted against the rights of the pregnant women. Although Cahill doubts that we can jettison the use of rights language altogether, if we are going to use rights language, she says, we must "remove that language from the context of moral and political liberalism" (361).

If we do so, we might be able to see that we have duties and obligations to which we do not explicitly consent. As Cahill puts it, "such obligations originate simply in the sorts of reciprocal relatedness that constitutes being a human" (361).6 For example, moving away from an individualistic liberal view of the pregnant woman as primarily or exclusively an autonomous moral agent might lead us to recognize the obligations that individuals and communities have to support her during and after a burdensome pregnancy (363).

We do not need to accept Cahill's commitment to the Catholic common good tradition to recognize the truth in her conclusion that pitting the rights of the fetus against the rights of the pregnant woman individualistically construed leads us to overlook important social dimensions of the problem of abortion. It seems to me that much the same dynamic is evident in the stem cell research debate.

Consider again the central argument that the Catholic

church has made against stem cell research. The Pontifical Academy for Life suggests that the fundamental ethical issue is whether it is morally licit to produce or use human embryos to derive embryonic stem cells. The reasoning the Academy provides for concluding it is not licit is worth reproducing in full. The Academy lists five points: 1. On the basis of a complete biological analysis, the living human embryo is—from the moment of the union of the gametes—a human subject with a well defined identity, which from that point begins its own coordinated, continuous and gradual development, such that at no later stage can it be considered as a simple mass of cells.

2. From this it follows that as a "human individual" it has the right to its own life; and therefore every intervention which is not in favor of the embryo is an act which violates that right. …

3. Therefore, the ablation of the inner cell mass (ICM) of the blastocyst, which critically and irremediably damages the human embryo, curtailing its development, is a gravely immoral act and consequently is gravely illicit.

4. No end believed to be good, such as the use of stem cells for the preparation of other differentiated cells to be used in what look to be promising therapeutic procedures, can justify an intervention of this kind. A good end does not make right an action which in itself is wrong.

5. For Catholics, this position is explicitly confirmed by the Magisterium of the Church which, in the Encyclical Evangelium Vitae, with reference to the Instruction Donum Vitae of the Congregation for the octrine of the Faith, affirms: "The Church has always taught and continues to teach that the result of human procreation, from the first moment of its existence, must be guaranteed that unconditional respect which is morally due to the human being in his or her totality and unity in body and spirit: The human being is to be respected and treated as a

person from the moment of conception; and therefore from that same moment his rights as a person must be recognized, among which in the first place is the inviolable right of every innocent human being to life'"(No. 60). (Pontifical Academy for Life, 2000; emphasis in original) Notice that the core of the argument, namely points one and two, is framed in terms of the rights of the individual embryo. We have seen this emphasis already in noting Richard Doerflinger's various statements on stem cell research. Yet, notice also the claim that we know the embryo to be an individual with rights on the basis of "a complete biological analysis." This is not, of course, the first time that the Catholic church has made this claim. In the Declaration on Procured Abortion, the Congregation for the Doctrine of the Faith claimed that "modern genetic science" confirms the view that "from the first instant, the programme is fixed as to what this living being will be: a man, this individual-man with his characteristic aspects already well determined" (Congregation, "Declaration on Procured Abortion,"1974, 13). The Instruction on reproductive technology, Donum Vitae, also makes this claim. "The conditions of science regarding the human embryo provide a valuable indication for discerning by the use of reason a personal presence at the moment of the first appearance of a human life: how could a human individual not be a human person?" (Donum Vitae, 13).

One reason the Catholic church has played such a major role in framing the stem cell debate is that, in defending its position, it combines the two claims we have just noted, neither of which is explicitly religious. First, the early embryo is an individual person with rights and, second, the fact that the embryo is an individual person is confirmed by modern science. Indeed, a fair amount of the literature that supports embryo research generally can be read as an attempt to answer the question posed in Donum Vitae: How can a human individual not be a human person? Certainly Catholic writers who reject the church's teaching on the status of the embryo have responded directly to that question (See Cahill, 1993; Farley, 2001; McCormick, 1994; Shannon, 2001; Shannon and Walter, 1990), but so too have non-Catholics. For example, in a statement issued by their ethics committee, what was then called the American Fertility Society rejected the claim in Donum Vitae that science supports the personhood of the embryo.

According to the ethics committee "... it remains fundamentally inconsistent to assign the status of human individual to the human zygote or early pre-embryo when compelling biological evidence demonstrates that individuation, even in a primitive biologic sense, is not yet established. Thus, homologues (identical) twins may result from spontaneous cleavage of the pre-embryo at some point after fertilization but prior to the completion of implantation. Furthermore, during very early development, an embryo is not clearly established and awaits the differentiation between the trophoblast and the embryoblast" (American Fertility Society, 1988, 3S).

Arguably, writers like Mary Anne Warren and Bonnie Steinbock, who distinguish between biological or genetic humanity and moral humanity, are also at least indirectly answering the question posed in Donum Vitae (Warren, 1997; Steinbock 2001, 1992). Yet, whether writers are responding more or less directly to Catholic discourse, or not at all, the important point is that the stem cell debate has been remarkably preoccupied with the question of whether the early embryo is an individual person and whether and how the minute details of embryological development help us to answer this question. This is one reason why a fair amount of the ethics literature on the topic reads like a textbook on embryology.

I want to be clear here: I am not suggesting that the details of embryological development are unimportant. The maxim from the field of research ethics applies here as well: bad science is bad ethics.

My point is rather that the preoccupation with the details of early embryogenesis may lock us even more rigidly into an individualistic human rights framework than we are in debates about abortion. It also leads us to frame the debate as fundamentally about one question, and, indeed, it tempts us to treat the question as if there is one and only one answer. In this frame of mind, once we have that answer, there is not a lot more to talk about. Either the early embryo is a person with the right to life, in which case embryonic stem cell research is wrong, or the early embryo is not a person with rights, and then there is no moral reason to object to stem cell work. Gene Outka has made a similar point in his assessment of stem cell literature. As Outka puts it, in its starkest form, the crystallizing question is whether it is cogent to claim that embryonic stem cell research is morally indistinguishable from murder (Outka, 84). The problem with framing the question this way, he says, is that it "encourages an unfortunate tendency to restrict evaluative possibilities to a single either/or. Either one judges abortion and the destruction of embryos to be transparent instances of treating fetuses and embryos as mere means to other's ends, or one judges abortion and embryonic stem cell research to be, in themselves, morally indifferent actions that should be evaluated solely in terms of the benefit they bring to others." (Outka, 2002, 184).

The frame of human rights reinforces this either/or because, as I noted, a being is either a rights-bearing entity or it is not. I have argued elsewhere, that this either/or tends to drive people to the extremes. Either the embryo is a person or it is essentially a kind of property (Lauritzen, 2001). Although I will not rehearse the argument for rejecting the two extremes here, it is worth noting that the rhetoric associated with each extreme does not appear to match the practice of those who adopt the rhetoric or in fact to match the considered moral judgments of most Americans on these issues.

I can illustrate my point in relation to the view that the

early embryo is a person with the right to life by describing a cartoon that hangs on my office door (See Figure A). The cartoon depicts protestors in front of a stem cell research lab condemning those who work there as being anti-life. Down the street at the abortion clinic, the workers are noting how quiet things have gotten at their facility since the stem cell lab opened. The point of the cartoon, of course, is that we may soon see protests and demonstrations of the sort that are common at abortion clinics at facilities that conduct stem cell research and that there is an irony in the fact that pro-life advocates would be demonstrating against research being done to find treatments for Alzheimer's, Parkinson's, and other devastating illnesses. This is not entirely fair to the pro-life community, but it makes a point.

In fact, I do not have trouble imaging protestors picketing stem cell research facilities for, as we just noted, when stem cell research and abortion are evaluated together and when the evaluative option is a single either/or, then abortion and stem cell research may appear indistinguishable from murder. Certainly the rhetoric of someone like Richard Doerflinger has been consistent in condemning both abortion and stem cell research as equivalent to murder. The cartoon draws attention to this consistency, even while it questions the commitment of pro-life advocates to scientific research designed to promote the quality of life.

In one sense, then, the cartoon probes whether there is an inconsistency between being pro-life and opposed to research an Alzheimer's disease, Parkinson's disease, and other devastating illnesses. I do not myself think that there is any inconsistency in being pro-life and opposed to stem cell research, but the cartoon does point in the direction of a fairly significant disconnect between the rhetoric and the reality of those opposed to stem cell research because they believe the early embryo is a person. To see this point, imagine that, instead of a stem cell clinic, the cartoon depicted on IVF clinic down the block from the abortion clinic and that the workers at the abortion clinic are noting how quiet things have gotten since the IVF clinic opened. The dramatic tension that made the original cartoon funny would be missing from our revised cartoon precisely because it is hard to imagine protestors disrupting the work at IVF clinics. To be sure, the Catholic church and others have argued that IVF is morally wrong, but the rhetoric condemning IVF is exceptionally muted compared to that condemning abortion or stem cell research. Nor has there been a concerted effort to put an end to IVF practice in this country as there has been in the case of abortion and stem cell research. Yet, if the embryo is a person from conception, then participating in IVF as it is practiced in this country, when early embryos are routinely frozen or discarded or both, is to be complicit with murder. Why, then, are there no organized efforts to shut down IVF clinics in this country?7 Indeed, opponents of stem cell research and cloning often write as if these technologies raise the haunting specter of human embryo research for the first time. The reality, of course, is that the existence of in vitro fertilization depended entirely on embryo research and that every variation or innovation in IVF protocols involves experimentation on human embryos. Carol Tauer is one of the few scholars who has pressed this point. As Tauer sees it: ...the entire history of the research leading to the first successful IVF is the history of attempts to fertilize oocytes in the laboratory. Eventually these attempts succeeded, and the first IVF baby was born, followed by thousands of others in the ensuing decades.

The ethics literature contains scholarly discussions as to whether it is ethically permissible to make use of medical advances that result from unethical research. This discussion sometimes focuses on medical research conducted by the Nazis in concentration camps and institutions for retarded, mentally ill, and handicapped persons. Yet I have never seen reference to reproductive technologies in this context. If the fertilization of embryos in research is a practice that is abhorrent to many or most people, then would it not be logical to question the continuing use of the results of such research? (Even the Catholic Church, which opposes the use of IVF and most other forms of assisted reproduction, does not invoke this argument to support its opposition.) (Tauer, 2001, 153) If the embryo research associated with IVF points to a problem of consistency for those who oppose stem cell research because it involves destroying persons, it is no less problematic for those who support stem cell research but insist on respecting the embryo and embrace the distinction between "research" and "spare" embryos.

For as Tauer points out, Robert Edwards, the scientist involved in the first successful IVF procedure, began studying fertilization nearly thirty years before Louise Brown was born in 1978, and the first successful laboratory fertilization of human eggs took place a full ten years before she was born. Tauer quotes Edwards' report on this work: "We fertilized many more eggs and were able to make detailed examinations of the successive stages of fertilization. We also took care to photograph everything because we would have to persuade colleagues of the truth of our discoveries" (Tauer, 154).8 Nor was the creation of these "research" embryos done secretly: Edwards and Steptoe published their work in the journal, Nature in 1970 (Edwards, Steptoe, and Purdy, 1970).

At the very least, then, there is something of an irony in the fact that so much attention has been devoted to developing and defending the distinctions between embryos created solely for research and embryos left over from IVF procedures, because there would be no embryos left over from IVF procedures had there not been embryos created solely for research purposes to develop IVF in the first place. Given this fact, and given that this fact is no great secret even though it has not been discussed very much it appears disingenuous to endorse the distinction between "research" and "spare" embryos as a way of demonstrating respect for the early embryo while nevertheless encouraging its destruction.

I have suggested that the fact that so much of the stem cell debate has been framed in terms of whether the embryo is a person with rights has been unfortunate because it has cast the debate in sharply individualistic terms and has led to a preoccupation with embryological development narrowly construed. In addition, however, framing the debate in terms of embryo status and embryo rights tends to exaggerate the differences among commentators in contrast to their similarities. Consider, for example, the response of conservative Judaism in the United States to this issue. Rabbi Elliot Dorff has prepared a responsum on stem cell research for the Rabbinical Assembly Committee on Jewish Law and Standards, and his responsum is instructive.9 As responsa are, it is structured in terms of relevant questions: in this case, two questions frame Dorff's discussion.

First, "may embryonic stem cells from frozen embryos originally created for purposes of procreation or embryonic germ cells from aborted fetuses be used for research?" (Dorff, 2002, 1). Second, "may embryonic stem cells from embryos created specifically for research, either by combining donated sperm and eggs in a petri dish or by cloning be used for research?" (1) I think it is noteworthy that the very questions that frame Dorff's analysis both reflect and perpetuate a certain construction of the issue, but at this juncture, my point is different: given the way the debate has been framed, what most (non-Jewish) readers of Dorff's analysis are likely to focus on is the difference between his treatment of the early embryo and that of others in the literature. Indeed, even where you might expect to find and do in fact find on closer inspection similarities between this Jewish analysis and Catholic reflection on this issue, the first impression will be that of difference.

The reason, of course, is that our attention is drawn to Dorff's analysis of the early embryo, and Jewish views are sharply different about embryo status than those of the Catholic church and other pro-life opponents of stem cell work.

For example, Dorff points out that, according to the Talmud, during the first forty days of gestation, the embryo and the fetus are considered as simply water. From the forty-first day until birth, Jewish tradition considers the fetus as "the thigh of its mother." Moreover, says Dorff: As it happens, modern science provides good evidence to support the Rabbis' understanding. As Rabbi Immanuel Jakobovits noted long ago, the Rabbis' "forty days" is, by our obstetrical count, approximately fifty-six days, for the Rabbis counted from the woman's first missed menstrual flow, while doctors today count from the point of conception, which is usually about two weeks earlier. By 56 days of gestation by obstetrical count the basic organs have already appeared in the fetus. Moreover, we now know that it is exactly at eight weeks of gestation that the fetus begins to get bone structure and therefore looks like something other than liquid.

Indeed, the Rabbis probably came to their conclusion about the stages of development of the fetus because early miscarriages indeed looked like "merely water," while those from 56 days on looks like a thigh with flesh and bones. (16) The contrast with Catholic teaching could hardly be more striking. Not only are Jewish views of the status of the early embryo notably different, but Jewish tradition claims scientific validation for its view of the embryo, just as Catholic tradition does. Not surprisingly, therefore, where Dorff answers both questions posed by the responsum in the affirmative, Catholic tradition would answer both negatively.

These differences are significant and must be attended to, but it is worth asking whether focusing on these differences does not obscure important similarities. Consider some of the similarities.

In sketching the Jewish view of stem cell research, Dorff notes that certain theological commitments are central. He lists at least three that would be strikingly similar to Catholic and other Christian theological commitments.

- Our bodies are not ours; they belong to God and God commands that we seek to preserve life and health.
- All human beings, regardless of ability or disability are created in the image of God and are therefore to be valued as such.
- Humans are not God. We are finite and fallible and this fact ought to promote humility and urge caution.

Now if we focused merely on questions of embryo status, we would miss entirely these similarities between Catholic and Jewish views. More importantly, we would miss the fact that these similarities may underwrite significant moral reflection on stem cell research that is not rooted in concerns about the early embryo.

For example, Dorff notes that, given Jewish theological and legal commitments, the provision of health care must be understood as a communal responsibility. Thus, access to therapies developed through stem cell research is a crucial issue of justice for the Jewish community. This theme is echoed in Laurie Zoloth's work on stem cell research. As Zoloth puts it: Research done always will mean research foregone.

Will this research help or avoid the problem of access to health, given that poverty and poor health are so desperately intertwined in this country?...How can difficult issues of global justice and fair distribution be handled in research involving private enterprise? (Zoloth, 2001, 238) Surely these are questions that any Catholic moral theologian would gladly press.

Indeed, attending to the similarities between Zoloth's work and Catholic reflection on stem cell research brings us back to Lisa Cahill's observation about debates on abortion: they tend to be too focused on questions of rights individually construed. When one shifts the frame of analysis, new and different issues and new and different ways to approach the same issues come into view. Notice, for example, how close Zoloth and Cahill are on the issue of rights.

According to Zoloth, Jewish tradition foregrounds questions of "obligations, duties, and just relationships to the other, rather than the protection of rights, privacy, or ownership of the autonomous self" (96). This leads Zoloth to ask: "Can the interests of the vulnerable be heard in our debate?" (105). To be sure, the American bishops have wanted to emphasize the vulnerability of the early embryo when they have asked this question, but Catholic tradition, like Jewish tradition, requires that we ask this question in a way that is not captured when moral emphasis is merely about individual rights and personal autonomy.

Or consider another shared sensibility that emerges if we move away from questions of embryo status, namely a wariness about the human tendency to hubris and overreaching. Zoloth put this point eloquently in relation to the biblical story of the Tower of Babel.

She notes a rabbinic midrash on this text: "when a worker was killed, no one wept, but when a brick fell, all wept." Zoloth comments on this midrash as follows: It was this decentering of the human and reification of the thing that was the catastrophe that felled the enterprise... It is not just that they breached a limit between what is appropriate to create and what is not, the process of the creation must be carefully mediated, with deep respect for persons over the temptations of the enterprise. Such a text elaborates on the tension between repairing the world ... and acts that claim that the world is ours to control utterly (Zoloth, 2001, 106-107).

Beyond Questions of Embryo Status This passage from Zoloth helps to illustrate the point I wish to make in arguing that the stem cell debate has been too focused on questions of embryo status and that we must move beyond status questions if we are fully to do justice to the moral questions raised by technological developments associated with stem cell work. For concern about human efforts utterly to control the world is not a moral worry narrowly tied to status questions.

Let me put this point in the form of a question that has not typically been asked in the stem cell debate: Is adult stem cell work as unproblematic as it is often assumed to be? That this is a productive question is suggested by testimony of Francis Collins before the President's Council on Bioethics in December 2002. Collins was asked to speak about the topic, "genetic enhancements: current and future prospects" and he specifically addressed the issue of pre-implantation genetic diagnosis (PGD). Although PGD is usually understood to involve screening IVF embryos and discarding unwanted ones, it is also possible to screen gametes. Because gamete screening may not have broad utility, Collins did not discuss the issue at length. He did, however, offer an interesting observation about gamete selection. Focusing on gametes, he says, is useful because it "isolates you away from some of the other compelling arguments about moral status of the embryo and allows a sort of cleaner discussion about what are the social goods or evils associated with broad alterations in the sex ratio and inequities in access to that technology" (Collins, 2002, 7). In other words, if in the future we could screen gametes in the same way that we can now screen embryos, most of the moral issues raised by PGD would apply to gamete screening, even though gametes are not embryos.

Might we not make a similar claim about embryonic and adult stem cell research? Do not many of the most pressing issues raised by embryonic stem cell technology remain when our focus is adult stem cell work rather than embryonic stem cell research? The fact that we do not immediately answer yes to this question, is testament to how decisively the debate about abortion has structured the stem cell debate. Nevertheless, we need to see that the answer to this question is yes and we need to see why.

Although I will not try to address all of the issues raised jointly by embryonic and adult stem cell research, it is worth highlighting several that I think require fuller discussion particularly with respect to adult stem cell work than they have yet received.

Commodification Issues Moral concerns about the commodification of gametes and embryos have been discussed extensively in the bioethics literature both in relation to reproductive technology and in connection with embryonic stem cell work (See Andrews and Nelkin, 2001; Annas, 1998; Corea, 1986; Radin, 1996; Resnik, 2002; Ryan, Ethics and Economics, 2001). Suzanne Holland, for example, has discussed the growing commodification of the human body in the biotech age. She cites a series of articles published in the Orange County Register that documents a vast for-profit market in human body tissue (Holland, 2001, 266). The Register's investigative reporters documented that most nonprofit tissue banks obtain tissue from cadavers donated by family members of the deceased for altruistic reasons. Most relatives are not told, and in fact have no idea, that donated body parts will be sold for profit. As Holland puts it: The "gift of life" is big business in America.

For a nonprofit tissue bank, one typical donation can yield between $14,000 and $34,000 in downstream sales, sometimes far more than that. "Skins, tendons, heart valves, veins, and corneas are listed at about $110,000. Add bone from the same body, and one cadaver can be worth about $220,000." Four of the largest nonprofit tissue banks told the Orange County Register that together they expected to produce sales totaling $261 million in 2000.

Nor is the issue of downstream commodification restricted to the sale of donated cadaveric tissue; it arises in relation to IVF embryos donated for research. As Dorothy Nelkin and Lori Andrews point out in their book, Body Bazaar, IVF patients are not generally told what the research involving their donated embryos will include.

Many will be unaware that their embryos will be used to develop commercial stem cell lines (Nelkin and Andrews, 2001, 35).

It is significant that even the most vocal advocates of procreative liberty and laissez-faire arrangements in reproductive matters recoil from the prospect of selling human embryos. Yet, although the commodification of tissue may be particularly troubling when it involves embryos, if there is a problem with commodifying and commercializing human tissue, it is a problem we confront with adult stem cell research as well as with hES cell work.10 Lori Knowles has made a similar point about being consistent in our moral judgments about commodifying embryos. She notes that fears about commodifying reproduction have led many to oppose the sale of embryos and to reject the idea that couples who donate embryos have any proprietary interest in the result of the research done with their embryos. As Knowles puts it, "if it is wrong to commercialize embryos because of their nature, then it is wrong for everyone. It is simply inconsistent to argue that couples should act altruistically because commercializing embryos is wrong, while permitting corporations and scientists to profit financially from cells derived by destroying those embryos" (Knowles, 1999, 40).

Knowles draws attention here to the fact that there is a tension between our moral and legal traditions as they apply to developments brought about by cloning, stem cell research, and the existence of ex utero human embryos, among other technological breakthroughs. For example, we patent embryonic stem cell lines, thereby insuring massive profits for patent holders, while decrying the commodification of embryonic life. Knowles is correct that there is a tension between our profit-based medical research model and our commitment to altruism and the access to health care that a commitment to justice demands. Adult stem cell research, of course, raises very similar issues, because the same tension exists between the need for proprietary control of technology and the need for affordable access. For example, patents have been sought and granted for human adult stem cells work as well as for embryonic stem cell technologies. According to a study on the patenting of inventions related to stem cell research commissioned by the European Group on Ethics in Science and New Technologies, as of October 2001, two thousand twenty-nine patents were applied or granted for stem cells and 512 patents were applied or granted for embryonic stem cell work (Van Overwalle, 2002, 23; see also McGee and Banger, 2002). As a result, access to therapies developed from adult stem cell research is likely to be as serious an issue as access to embryonic stem cell therapies.

Indeed, issues arising from commodification of both adult and embryonic stem cells are likely to dominate the next phase of the debate, if only because corporate interest in this work, both nationally and internationally, is so strong (Hill, 2003). Noting that, as of 2002, "a dozen biotech companies have entered the stem cell industry and have invested millions of dollars," David Resnik suggests that the next stage of the stem cell debate will involve a battle over property rights relating to stem cells (Resnik, 2002, 130- 31).

To be sure, the battle over property rights raises important legal and policy question, but it also raises ethical questions as well. For example, Resnik provides the following table of possible ethical objections to patenting stem cells.

Although the ethical objections to commodifying stem cell work can not be sorted out as neatly as this table suggests, Resnik correctly identifies a variety of such objections. I will not try to defend it here, but my own view is that the problem is not just that stem cells and their products may be commodified, but that market rhetoric may come to dominate the discussion (and practice) of regenerative medicine in a way that is dehumanizing. That is, *Deontological Objections to Property Rights in ES cells Consequentialist Objections to Property Rights in ES cells* ES cells (and their products) should not be treated as property; they should be viewed as having inherent dignity or respect.

Treating ES cells as property will have dire social consequences, such as exploitation, destruction of altruism, and loss of respect for the value of human life.

ES cells should not be treated as private property; they are *res communis* or common property.

Treating ES cells as property will undermine scientific discovery and technological innovation in the field of regenerative medicine. (Resnik, 2002, 139) market-rhetoric may lead us ultimately to think of humans as artifacts. In short, I take very seriously Margaret Radin's argument that the rhetoric in which we conceive our world affects who and what we are (Radin, 1996, 82).11 At the same time, however, stem cell research is most likely to bear therapeutic fruit, if there is a market in stem cells and their products. The pressing moral question, then, is how do we promote the benefits that stem cell research may yield without succumbing to a market rhetoric that reduces humans to commodities? Several writers have suggested that one answer may be to promote greater governmental regulation. For example, Holland argues for moving beyond a policy of restricting federal funding of stem cell research but allowing an unregulated private market in this field to active regulation to curb the private sector's work on stem cells. Lori Knowles suggests that the United States might adopt a body like Canada's Patented Medicine Prices Review Board as a way to allow a market to function, but with oversight that would provide access to potential cells for further research and price controls of products to insure widespread access (Knowles, 1990, 40). George Annas has suggested that we need to establish a federal Human Experimentation Agency to regulate in the area of human experimentation. (Annas, 1998, 18). As Annas puts it: Virtually all those who have studied the matter have concluded that a broad-based public panel is needed to oversee human experimentation in the areas of genetic engineering, human reproduction, xenografts, artificial organs, and other boundarycrossing experiments. (19) Francis Fukuyama has argued that a new agency with a mandate to regulate biotechnology on broad grounds and in both the public and private sector may be needed (Fukuyama, 2002, 215). Vanessa Kuhn argues that "it is time to put in place legislation that will deter stakeholders from licensing their technology to one exclusive distributor and thus creating a monopoly market, which would set artificially-high prices and lead to less access for the sick especially for the uninsured, the poor, and the elderly" (Kuhn, 2002). To that end, Kuhn identifies four possibilities: • Development of a new kind of patent.

- Set limits on exclusive licensing through compulsory licensing.
- Lower the lifespan of hES cell patents
- Set stricter guidelines for hES patent utility

I do not have the expertise to make policy recommendations, but let me stress two points. First, the policy issues with regard to commodifiying adult stem cell work will be as vexing as those confronting regulation of embryonic stem cells. Second, although these questions may at first appear to be strictly legal or largely political matters, they involve serious value judgments about the common good that are every bit as morally vital as questions about the status of the embryo. I thus agree with Gene Outka, that not to confront directly questions about how stem cell research will be organized, financed, and overseen is a kind of ethical failure (Outka, 2002, 177). Obviously, for example, the institutional arrangements for conducting stem cell research have implications for the questions of justice we previously noted. The fact that so much stem cell research is being done by private corporations insures future conflict.

On the one hand, corporations have fiduciary obligations to their shareholders and will therefore seek to control access to stem cell lines or therapies developed from those lines through patent protection and licensing agreements. On the other hand, such a system is likely to further widen the gap between the health care haves and have-nots (See Lebacqz 2001; McLean 2001).

Moreover, as Karen Lebacqz notes, if justice is an important consideration in deliberations about stem cell work, then it ought to shape the research agenda. The example she gives to make this point is worth noting. Just as with organ transplants, tissue rejection may be a major problem for stem cell therapies. This is one reason that the prospect of combining stem cell work with somatic cell nuclear transfer (SCNT) has been so enticing. With this combination, you could in theory develop tissue that would be completely histocompatible. Nevertheless, according to Lebacqz, developing stem cell therapies with SCNT is "highly questionable," if justice is a primary consideration. The reason is that unique cell lines would need to be created for each patient, and that is likely to be very expensive and thus unaffordable for many. Although it would certainly be less expensive, the same would likely be true for adult stem work. For that reason, rather than pursuing an individualized approach to stem cell research, concerns about affordable access to new therapies might urge the pursuit of universal donor cell lines.

Embodiment, Boundary Issues, and Human Nature In discussing the debate about embryo status, I focused primarily on the contested question of whether the early embryo is a person with the right to life. We saw that this question tends to lead to the mobilization of minute details of embryological development to support one's view of the embryo. Yet, if attention to embryo status tends to focus us on the microscopic, viewing stem cell research through the lens of embryological development can also have a kind of telescopic function through which larger issues come into view.12 For example, Catherine Waldby and Susan Squier argue in a forthcoming issue of the journal Configurations that focusing on stem cells and embryonic life leads us fundamentally to question what it means to be human. According to Waldby and Squier: Stem cell technologies have profound temporal implications for the human life course, because they can potentially utilize the earliest moments of ontogenesis to produce therapeutic tissues to augment deficiencies in aging bodies. Hence they may effect a major redistribution of tissue vitality from the first moments of life to the end of life. In doing so however they demonstrate the perfect contingency of any relationship between embryo and person, the non-teleological nature of the embryo's developmental pathways. They show that the embryo's life is not proto-human, and that the biology and biography of human life cannot be read backwards into its moment of origin. (Waldby and Squier, forthcoming) The claim that there is a perfect contingency in the relationship between embryo and person may at first appear to be just another "microscopic" claim about embryo status, but it is clear that Waldby and Squier mean to imply much more in asserting that the embryo's development is non-teleological. In effect, they reject the notion that there is a meaningful trajectory to human life. What was killed, they say, when stem cells were first derived from the inner cell mass of a blastocyte was not a person, but a "biographical idea of human life, where the narrative arc that describes identity across time has been extended to include the earliest moments of ontogeny" (Waldby and Squier, forthcoming).

That much more is at stake here than the question of whether the embryo is a person is clear if we attend to the notion of a trajectory of a human life. Gilbert Meilaender, for example, has argued that our attitudes toward death and dying are importantly shaped by our conception of what it means to have a life (Meilaender 1993). Indeed, according to Meilaender, two views of what it means to have a life of what it means to be a person have been at war with each other within the field of bioethics over the past thirty years and these views underwrite sharply different views not just about the issues of abortion or euthanasia that are implicated here but with regard to practically every moral issue we might confront in the field of bioethics.

On Meilaender's view, having a life means precisely that there is a trajectory that traces a "natural pattern" in embodied life that "moves through youth and adulthood toward old age and, finally, decline and death". As he puts it elsewhere in this essay, "to have a life is to be terra animata, a living body whose natural history has a trajectory".

Although Meilaender develops the notion of a natural trajectory of bodily life primarily to address the issue of euthanasia and not stem cell research, his talk of "natural history," "natural pattern," and "natural trajectory" draws attention to one of the most significant issues raised by stem cell research and related technologies. Does stem cell research undermine the very notion of a human life constrained by natural bodily existence? The example on which Meilaender focuses here is instructive for thinking about the broad implications of stem cell research in this regard. If stem cell therapies fundamentally alter our sense of a natural pattern to aging would they not also fundamentally alter our sense of what it means to be human? Meilaender's answer is that such a change would fundamentally affect what it means to be human precisely because we are embodied creatures and for that reason our identity is tied to the body and the body's history.

Leon Kass has made a similar point recently in reflecting on the prospect that regenerative medicine might significantly lengthen the human life span (Kass, 2003). He, too, invokes the notion of a natural trajectory, one that stem cell research may undermine.

Although it is possible to approach the prospect of extending the human life-span in an abstract way, he says, to think of what such a change would mean experientially is to recognize that "the 'lived time' of our natural lives has a trajectory and a shape, its meaning derived in part from the fact that we live as links in the chain of generations" (13). Indeed, says Kass, without something like the natural trajectory of bodily life that currently exists, the relationship between the generations would be decidedly different, and probably not better. "A world of longevity," writes Kass, "is increasingly a world hostile to children" (13). Walter Glannon has argued that, at the very least, increased longevity would increase competition for scarce resources between older and younger generations. According to Glannon, "it is at least intuitively plausible that an over populated world with substantially extended human lives and scarce resources could adversely affect the survival and reproductive prospects of the young and harm them by thwarting their interest in being healthy enough so that they could survive and procreate" (Glannon, "Extending," 347).13 Francis Fukuyama has also suggested some of the reasons why increased longevity may imperil children, but he also notes that our relationship to death may change as well (See 2002, ch. 4). "Death," he says, "may come to be seen not as a natural and inevitable aspect of life, but a preventable evil like polio or measles. If so, then accepting death will appear to be a foolish choice, not something to be faced with dignity or nobility" (Fukuyama, 2002, 71).

Sometimes the question of the transformative possibilities that come with stem cell research is raised even more starkly when the question asked is not how may stem cell work affect what it means to be human, but instead: Does stem cell research open the door to a post human future? This is a point Waldby and Squier raise explicitly when they discuss the combination of genetic engineering and stem cell therapy. They suggest, for example, that xenotransplantation forces us to confront the prospect of transgressing species boundaries.14 The conclusion of their paper is worth quoting in full: Thus the ontological status of the embryo is not the only thing in question. The ontological status of the graft recipient must be negotiated, when the graft involves genetically-engineered stem cells from another species. And the ontological status of the illnesses to which biomedical technology responds is equally challenged, in an endless regression, as the division between veterinary and human medicine, or between zoonoses (diseases humans can catch from animals) and what has recently been dubbed humanooses, is called into question. This increasingly permeable, increasingly constructed barrier between human and animal presents us with another form of life to negotiate, whose boundary lies not between silicon and carbon, but rather between steps in the evolutionary ladder or the branching development tree of phylogenetic lifeforms. Stem cell technologies thus challenge both the temporal and spatial boundaries of human life, both our biography and our biological niche, giving a much broader meaning to the questioning of embryonic personhood. (Waldby and Squier, forthcoming) Regrettably, with some notable exceptions, the ethical debate about stem cell research has not taken up in a sustained way what it would mean to pursue stem cell therapies that might significantly undermine the notion of a natural human life or erode the boundary between human and non-human species. When the issue is framed in terms of the status of the embryo, the question tends to be whether the research should be conducted at all. By contrast, when the issue is framed in terms of adult stem cell work, the question is not whether, but how and with what consequences. Yet, that is a question we have not systematically answered. Given the potential for good embedded in the prospects of adult stem cell research, it is not surprising that there appears to be widespread and largely uncritical acceptance of adult stem cell research. But, if the promise of stem research is as revolutionary as is often claimed, we are going to need a much more expansive discussion of stem cell research both embryonic and adult than we have had heretofore. Obviously, I cannot explore this more expansive horizon in any detail in this report, but let me in closing suggest one direction we need to explore.

To signal the decisive break that I think we may need from the usual bioethics frame, I want to draw attention to Martha Nussbaum's recent article in the journal Daedalus entitled, "Compassion & Terror" (Nussbaum, 2003). Discussing Euripides' play, Trojan Women, Nussbaum reflects on the fact that the Greek poets returned obsessively to the sacking of Troy and the acts of the "rapacious and murderous Greeks." She explores the poets' compassionate imagining of the fate of Trojan women and children to reflect on the conditions and limits of a compassionate vision.

Although Nussbaum is ultimately concerned about engendering a compassionate vision for Americans in the face of terror and particularly compassion for innocent women and children far from our shores, her analysis of compassion is thought-provoking in relation to stem cell research.

Nussbaum notes that compassion is a complex emotion requiring a series of judgments involving another person's suffering or lack of well-being. We must judge that someone has been harmed, that the harm is serious, and that it was not deserved. Moreover, says Nussbaum, Western tradition has stressed what could be called the "judgment of similar possibilities." In other words, "we have compassion only insofar or we believe that the suffering person shares vulnerabilities and possibilities with us" (Nussbaum, 2003, 15).

Now surely in just about everyone's catalogue of human vulnerabilities are illness, old age, and death. Yet, as we have just seen, stem cell research might significantly transform the "human" experience of illness and death, at least for some. If stem cell therapies were to erode the notion of human nature or species membership, might they not also erode some basic moral sensibilities? Mary Midgley, for example, has argued that both the notion of human nature and that of human rights are importantly tied to membership in our species because rights are "supposed to guarantee the kind of life that all specimens of Homo sapiens need" (Midgley, 2000, 9).

Although Nussbaum avoids the language of human nature, it is precisely this sort of point that she highlights when she argues that compassion requires the belief that others share vulnerabilities and possibilities with us. Indeed, like Midgley, Nussbaum ties the notion of universal human rights to important human functions and capabilities. The basic idea, she says, is to ask what constitutes the characteristic activities of human beings. In other words: 'What does the human being do, characteristically, as suchand not, say, as a member of a particular group, or a particular local community?' To put it another way, what are the forms of activity, of doing and being, that constitute the human form of life and distinguish it from other actual or imaginable forms of life, such as the lives of animals and plants, or, on the other hand, of immortal gods as imagined in myths and legends (which frequently have precisely the function of delimiting the human)? (Nussbaum, 1995, 72) Nussbaum notes that this inquiry proceeds by examining a wide variety of self-interpretations and that comparing characteristic human activities with non-human activities and, through myths and stories, comparing humans and the gods is particularly helpful.. For one thing, such an inquiry helps us to define limits that derive from membership in the world of nature.

Indeed, although Nussbaum is particularly attentive to the wide variety of cultural interpretations of what it means to be human, she insists that to ground any essentialist or universal notion of human rights, one must attend to human biology. Although her account of the human is neither ahistorical nor a priori, it is linked to an "empirical study of a species-specific form of life" (1995, 75).

When she develops her account of central human capabilities, she begins with the body. She writes: We live all our lives in bodies of a certain sort, whose possibilities and vulnerabilities do not as such belong to one human society rather than another.

These bodies, similar far more than dissimilar (given the enormous range of possibilities) are our homes, so to speak, opening certain options and denying others, giving us certain needs and also certain possibilities for excellence. The fact that any given human being might have lived anywhere and belonged to any culture is a great part of what grounds our mutual recognitions; this fact, in turn, has a great deal to do with the general humanness of the body, its great distinctness from other bodies.

The experience of the body is culturally shaped, to be sure; the importance we ascribe to its various functions is also culturally shaped. But the body itself, not culturally variant in its nutritional and other related requirements, sets limits on what can be experienced and valued, ensuring a great deal of overlap. (Nussbaum, 1995, 76) Nussbaum's work both in identifying the judgments that underwrite compassion and in tying an account of rights to human function and capabilities that are presumably universal highlights what is at stake, not merely with stem cell research but with a growing list of biotechnological developments which appear to destabilize the concept of human nature and which require that we think carefully and hard about what it might mean for some humans to have access to these technologies while other humans do not.15 At the very least, the combination of what Rabinow describes as the biologicalization of identity around genetics rather than gender and race with the possibility of manipulating that genetic identity for those with the money or power to do so does not bode well for securing wide-spread compassion across economic or technological divides. Even more important, however, is the recognition that the very notion of human rights may ultimately rest on the idea (and what, until recently has always been the reality) of a natural human condition that is relatively stable.

I believe that Nussbaum is correct when she claims that inquiring into characteristic human activities and comparing these to non-human activities helps us to define limits and thereby to promote human flourishing. Unfortunately, what Susan Squier calls the "pluripotent rhetoric" of stem cell research is that of limitless possibilities (Squier, Liminal Lives). The ultimate limit, of course, is death and yet even this limit appears illusory in some visions of our biotech future.

It is worth noting in closing that William Safire's New Year's Day column at the dawn of the twenty-first century, in January 2000, was entitled "Why Die?" The longing behind this question is neither new nor unfamiliar. What is new is that this longing to escape the vulnerabilities and limitations of the body is united with a technology that holds out the prospect of fundamentally changing that body.

Yet, I agree with Gerald McKenny that we need to ask whether we wish to accept and promote a view of bodily vulnerability as merely an obstacle to human flourishing, which ought to be overcome at any cost (McKenny, 1998, 223).

Although a longing for invulnerability is perhaps a quintessentially human trait, and although the quest to reduce the human suffering wrought by illness and disease is morally admirable, there is no mistaking the hubris behind the question, Why Die? Opponents of stem cell research have, from the start, argued that there is a kind of idolatry in a science that would reduce the human embryo to just so much biological material.

What I have tried to show in this report is that such concerns need not be limited to those who think that the early embryo is fully a person. Nor should this kind of concern be limited to those who oppose stem cell research. I do not think the early embryo is a person and I believe that both embryonic and adult stem cell research should go forward under a system of strict regulation. Nevertheless, I confess to being haunted by the passages with which I began and I believe that future debates on stem cell research must take very seriously the worries about commodification and the possibility of fundamentally changing the trajectory of a human life. Compare, for example, the various statements that Richard M. Doerflinger has made on the U. S. bishops' behalf. See Doerflinger, 1988, 1998, 1999, 2001.

Margaret Farley has argued that the Catholic preoccupation with abortion has eroded its credibility on other important social issues, including stem cell research. (See Farley, 2000). Another debate that is at least partly shaped by focusing on embryo status revolves around the question of complicity. For example, supporters of stem cell research may harbor a residual uneasiness about endorsing the destruction of human embryos, at least if the number of articles in the literature explaining the concept of complicity with wrongdoing is any indication. John Robertson, Ronald Green, and Thomas Shannon, have all written on the issue of cooperation with evil in relation to stem cell research. (Robertson, 1999; Green, 2002; Shannon, 2001; see also Kaveny, 2000; and Gilliam, 1997). To be sure, the issue of complicity or cooperation with wrongdoing is a very traditional one in moral philosophy and theological ethics.

Still, if the early embryo does not deserve the respect accorded persons and if destroying the embryo is compatible with respecting it, then deriving stem cells is not an act of wrongdoing and issues of complicity do not arise. Cahill also emphasizes the way a liberal individualist view of the person discounts the significance of embodiment. I will return to this point below. That both those who view the embryo as a person and those who do not but who insist on respect for the embryo, have been remarkably cavalier with regard to the use of embryos in IVF programs can be seen by the fact that there are currently over 400,000 embryos frozen in the United States, a number we did not even know until quite recently (Hoffman et al. in association with The Society for Assisted Reproductive Technology and RAND, 2003).

In a commentary published in Nature, in September 2001, Edwards writes: "On the verge of clinical application, stem cells offer a startlingly fundamental approach to alleviating severe incurable human maladies. Fondly believed to be a recent development, they have in fact been part and parcel of human in-vitro fertilization (IVF) from as long ago as 1962."

Dorff's responsum was accepted by the Committee on Law and Standards by a vote of twenty-two to one in March 2002. On the basis of Dorff's responsum the Rabbinical Assembly passed a resolution in April 2003 supporting stem cell research for therapeutic purposes. (Resolution in Support of Stem Cell Research and Education, April 2003: available at www.rabassembly.org.) Alpers and Lo draw the distinction between commodification and commercialization as follows: "The issue of commodification involves treating either human beings or symbols of human life as merchandise or vendible goods. ...

Commercialization refers to the practice of realizing large profits from the development and sale of techniques or products that involve distinctive human material, such as embryos, eggs, or tissue" (Alpers and Lo, 1995). Radin quotes Georg Lukács on the reification of commodities and the effects on human consciousness. Lukács writes: "The transformation of the commodity relation into a thing of 'ghostly objectivity' cannot therefore content itself with reduction of all objects for the gratification of human needs to commodities. It stamps its imprint upon the whole consciousness of man; his qualities and abilities are no longer an organic part of his personality, they are things which he can 'own' or 'dispose of' like the various objects of the external world" (Radin, 1996, 82). When we think about genes for enhancing memory or muscle mass, it is worth keeping in mind Lukács's claim that human qualities and abilities may come to be thought of as objects for sale in the external world. Erik Parens has noted the importance of attending to the big picture raised by stem cell work and how the politics of abortion has obscured that picture. See Parens, 2000. In another essay, Glannon argues that substantially increasing the human life span would profoundly affect issues of personal identity and thus a sense of personal responsibility for one's action. He ties his argument in interesting ways to the biology of memory function (Glannon, "Identity"). For a classic philosophical discussion of the problems associated with immortality, see Williams, 1973.

Although he is not discussing stem cell research explicitly, Paul Rabinow's discussion of technological change wrought during the last two decades is worth noting. He writes: "In the United States, for example, in the last two decades, while the most passionate value conflicts have raged around abortion, a general reshaping of the sites of production of knowledge has been occurring. To cite the biotechnology industry, the growing stock of genomic information, and the simple but versatile and potent manipulative tools (exemplified by the polymerase chain reaction) is to name a few key elements; a more complete list would include the reshaping of American universities, the incessant acceleration in the computer domains, and the rise of 'biosociality' as a prime locus of identity a biologicalization of identity different from the older biological categories of the West (gender, age, race) in that it is understood as inherently manipulable and re-formable" (Rabinow, 1999, 13).

A couple of pages later, he writes: "My analysis points to the fact that the basic understanding and practices of 'bare life' have been altered. The genome projects (human, plant, animal, microorganismic) are demonstrating a powerful approach to life's constituent matter. It is now known that DNA is universal among living beings. It is now known that DNA is extremely manipulable. One consequence among many others is that the boundaries between species need to be rethought; transgenic animals made neither by God nor by the long-term processes of evolution now exist (16). For a science fiction exploration of this theme of selected genetic enhancement, species boundary crossing see Octavia Butler, Dawn. A number of people either helped with the preparation of this report or provided feedback on an earlier draft. Thanks to Christa Adams, Diana Fritz Cates, William FitzPatrick, James L. Lissemore, Charlie Ponyik, Mary Jane Ponyik, Kristie Varga, Lisa Wells, and the ethics writers group at John Carroll University.

Women and Biotechnology

Today, women are faced with a rapidly expanding array of reproductive technologies. Developed by private biotechnology companies and marketed to fertility clinics, these new options have been presented to women as an issue of "choice"—supposedly providing them greater control over the process and outcome of their pregnancies. The Council for Responsible Genetics unequivocally supports a woman's right to make her own reproductive decisions. However, we oppose the utilization of human eggs and embryos for experimental manipulations and as items of commerce because of the potential for eugenic applications and health risks to women and their offspring. The realm of assisted reproduction has become a multi-billion dollar industry, visible in the increasing availability of vitro fertilization, prenatal genetic diagnosis, and chemical and chromosomal testing of the fetus. So far, these experimental procedures have not been closely regulated. Part of this is because the reach of federal oversight extends exclusively to publicly- funded research, leaving private sector activity largely unregulated. Many are resistant to any regulations over the fertility industry due to the continuing political residue of the abortion debate. Women in the United States fought hard to have courts recognize a women's right to choose abortion, and are wary of opening the door to government regulation of anything to do with women's eggs, embryos or fetuses. The word "choice" has powerful connotations and women's groups themselves are not eager to restrict its power. The industry is aware of this and has capitalized on the situation by supposedly offering the new genetic technologies as "choices." The unique role of women in reproduction puts them on the front line of bio-technological experimentation, and as such, women have the potential to play a leading role in determining the direction and scope of these developments. CRG works, in partnership with a number of women's health groups, to reshape the discussion about genetic technologies in reproduction and to equip women with the information necessary to lead the process. Our goal is to shift debates over genetic technologies away from abortion politics and into more effective and productive discussions about the integrity of reproduction and the control of women's bodies. Through education, networking, outreach, and activism women can direct their own reproductive futures.

Ancient Greek mythology is replete with references to part-human animals. There is the monstrous half-human, half-bull Minotaur; the Gorgon sisters (one of whom is Medusa) with hair of writhing snakes; the Sirens who are sweet singing sea nymphs each with the head of a woman and the body of a bird; and, not to be forgotten, there is the infamous Sphinx with the head and breasts of a woman, the body of a lion, and the wings of a bird. Part-human creatures are also a staple of modern science fiction, as in

H.G. Wells' *The Island of Dr Moreau*, where animals are vivisected into part-human creatures, or George Langelaan's *The Fly*, in which a scientist emerges from his disintegrator-reintegrator machine with the head and arms of a fly. But part-human animals are not only science fiction—they are also science fact. While not as monstrous as the creatures of lore, part-human laboratory animals raise some important ethical and societal issues.

2

Biotechnology: Risks and Safety Options

Safety of Biotech Crops

Many people who object to biotech crops argue that the crops should not be allowed to grow in the environment until science *proves* that they are safe. Others who support biotech crops argue that science has *proven* that they are indeed safe. However, the concept of proof has no place in science. Many people who do not actively practice science do not understand that science is structured so that scientists can never prove anything.

To understand why science cannot prove anything, we need to define and discuss two terms, "Science Literacy" and "Scientific Literacy." "Science literacy" is knowledge about science facts. Facts such as the Earth orbits the sun, photosynthesis takes place in plant cells, and atoms are made of electrons, protons, and neutrons are all examples of science literacy. But science literacy is very different from scientific literacy. "Scientific literacy" is knowledge of science as a *process* for understanding the physical world. So, what do we mean by "science as a process?" In other words, how does science work? Science and scientists utilize the scientific method, a standard, rigorous, and step-wise process to understand the world. After an idea or observation, scientists pose a hypothesis, which is a tentative explanation or assumption about some particular aspect of nature. To be valid, the hypothesis must be testable, and, more importantly, it must be refutable. When the hypothesis is tested, it is either supported or refuted. If the hypothesis is refuted, then an alternative hypothesis is posed and the process continues. If a hypothesis is not refutable (or falsifiable), it does not meet the minimum requirements of the scientific method. For example, a hypothesis that God created the earth cannot be refuted, and therefore it cannot be evaluated using the scientific method. Hypotheses and theories can never be proven true using the scientific method. Therefore, science advances only through disproof. This is a critical and often misunderstood point. To be scientific, theories can never be proven true, but all theories must be refutable. Therefore, all theories, and by extension all of science, are tentative. As an example, let's use a science fact that is known to most adults: the existence of electrons. We know that electrons exist, but here's the rub: Science can never prove that electrons exist. Hypotheses about the existence of electrons have been supported after countless tests using the scientific method. In other words, they have not been refuted. Knowledge of the precise nature of electrons will always be undergoing refinement, but the weight of scientific evidence clearly supports the existence of electrons.

Scientific Proof and the Debate Between Evolution and Creationism

Understanding how science works also is critical to understanding the seemingly never-ending debate between those who support evolution and those who support creationism. The debate typically devolves (no pun intended!) into the argument by creationists that scientists cannot prove that evolution has occurred; therefore, their hypotheses of creationism are equally valid. Hypotheses regarding numerous aspects of evolution have been posed and tested. As testing has winnowed out incorrect ones, our understanding has become ever more clear. As with electrons, knowledge of the precise nature of evolution will undergo constant refinement, but the total current weight of scientific evidence supports the existence of evolution. As scientists, we do not arrive at that conclusion out of capriciousness or because of a "leap of faith," but rather because of the systematic process for understanding physical reality called the scientific method. In fact, anyone who supports evolution because of faith or belief is not using scientific rationality. Several arguments posed by creationists do constitute hypotheses and therefore are falsifiable. The hypothesis that the Earth is no more than 10,000 years old has been proposed numerous times, but refuted in many ways. Conversely, alternative hypotheses that the Earth is much older have been repeatedly supported. Indeed, today the weight of evidence overwhelmingly suggests that the Earth is billions of years old. Unfortunately, many other arguments posed by creationists are not falsifiable and therefore cannot be addressed using scientific inquiry. How can we falsify the hypothesis that God or an Intelligent Designer created the major forms of life on Earth? Creationist theories that cannot be refuted represent a particular world view, but they are not science. These theories can be taught in schools as philosophy or religion, but they must not be taught as science, unless the instructor uses them to demonstrate to the class what is and what is not science.

Class Exercise

As the instructor, hold a rock three to four feet above the floor. Ask the students what will happen if you let go of the rock. Hopefully, they will say that the rock will drop to the floor! Then, ask the students to prove that the rock will fall to the floor before you let go. They invariably will argue that rocks always fall to the ground when released, so when you release the rock it also will fall. But their statement does not prove that the rock will fall (or prove the existence of gravity). Their statement is based on countless observations of objects falling to the ground when dropped from above the ground. They are basing their statement not on proof, but on an overwhelming weight-of-evidence of what has happened to objects after they were released. Scientific explanations of falling objects started with a hypothesis that the rock would fall to the ground when released. Remember, to be a valid hypothesis, the hypothesis must be refutable. In other words, the test of the hypothesis must allow for the observation that the rock will not fall to the ground when released. Like electrons, no one has ever seen gravity, but because of the scientific method of supporting or refuting hypotheses about gravity, there is overwhelming evidence that gravity exists.

How about another example? This time we'll use an example from plant biology and agriculture. A scientist states a hypothesis that adding nitrogen to the soil will result in increased grain production in corn (maize) plants. The scientist tests the hypothesis in a carefully controlled experiment. Her hypothesis is that nitrogen will increase grain production, and because the hypothesis must be subject to refutation, her alternative hypothesis is that nitrogen will not increase grain production. The experiment reveals that nitrogen does indeed increase grain production. Therefore, her initial hypothesis (also known as a null hypothesis) is supported. If the experiment had not resulted in increased grain production, the initial hypothesis would have been refuted and the alternative hypothesis would have been supported. The scientist can never prove that adding nitrogen to soil increases grain production, but if the hypothesis is supported time and time again, the weight-of-evidence convinces us that the relationship between nitrogen and increased grain production exists and is predictable. Now if we turn to the safety assessment process for biotech crops, we can see more clearly why a scientist using the scientific method can never prove that the crops are safe. To test hypotheses about the risk of biotech crops to human health, the hypotheses must be subject to testing and refutation. Remember, hypotheses can not be proven true, but they can be disproved. In the United States, determining whether human health risks from consuming foods derived from biotech crops are acceptable or unacceptable is the primary responsibility of the Food and Drug Administration (FDA). The determination of "is it safe enough" is a societal decision that involves much more than science. However, science and the scientific method do form the foundation upon which the decision is made. For crops produced using recombinant DNA technology, the FDA evaluates the new protein that has been introduced into the crop. Proteins that typically are considered safe (or that pose negligible risk) have the following characteristics:

- They are from sources with no history of toxicity or allergy
- They do not resemble known toxins or allergens
- They have well understood functions
- They are expressed at low levels (major allergens are usually found in large amounts in allergenic foods)
- They are rapidly degraded in the stomach (within minutes)
- There is a lack of adverse effects in mice at high levels of consumption
- Animal feeding studies (42-day poultry trial) do not reveal any adverse effects

To evaluate the characteristics described above, the FDA regulator evaluates data from the following studies:

- Acute oral toxicity in mice (Is the protein toxic when ingested by a mouse?
- In vitro digestibility under simulated gastric conditions (How quickly does the protein break down in conditions that mimic the human stomach?)
- Comparison of amino acid sequence similarity of the protein to known protein allergens (Is the protein structurally similar to proteins that we know are allergenic?)
- Thermolability of the protein (Does heat, such as we might see in food processing or cooking, break down the protein?)
- Protein expression in plant tissues (Where in the plant does the protein occur and in what amounts?)
- Protein fate in parts of the plant and in processed food fractions (How do the amounts of protein in different parts of the plant change over time and how much protein is present as the harvested crop is being processed into food?)
- Poultry feeding studies (Are there any differences between chickens fed the biotech crop and those fed a non-biotech crop over a 42-day period? This type of study gives the regulator some idea of potential long-term dietary risks associated with humans eating food with ingredients from biotech crops.)

What is important to recognize here is that none of the results from the studies alone or in combination prove anything. In particular, they do not prove that human health risks from biotech crops are acceptable or that they are safe. Each study tests a hypothesis. For example, the acute oral toxicity in mice study is centered around the initial hypothesis that the dose or doses of protein administered to the mice will not result in mortality or any signs of toxicity. The results from all of the studies are evaluated by the FDA regulator, who makes a decision about the food safety of the biotech crop. The regulator, therefore, utilizes a weight-of-evidence approach when making his decision. The results either provide a weight-of-evidence that the protein is safe to consume or that it is not safe to consume.

Biotech Crops Evaluated for Environmental Safety in the United States

First generation biotech crops (herbicide and insect resistant plants) have been in commercial use since 1995 in the United States. Farmers are adopting this technology readily because of two primary reasons: (1) it simplifies weed or insect management in their agricultural production systems, and (2) the economic benefits outweigh the costs. A common concern with the release of any biotechnology derived product is the effects it could have on the environment. In the previous chapter titled, "Is this Stuff Safe to Eat?," the regulatory agency we focused on was the FDA (United States Food and Drug Administration). When deciding whether or not a genetically engineered product is safe for the environment, the primary regulating agencies involved are the United States Department of Agriculture (USDA) and the United States Environmental Protection Agency (EPA). All three agencies (USDA, FDA, and EPA) overlap in their responsibilities, but each also has unique duties.

The Roles of the USDA and the EPA

In the United States, when determining if a food crop is safe enough to be grown in the environment the USDA and EPA are the most involved agencies. The USDA is a regulatory agency that monitors and regulates the use of biotechnology for agricultural purposes. They have the responsibility of reviewing data generated from field trials to determine whether a product or crop would have the potential to become a plant pest or cause any detrimental effects on the environment. Genetically engineered herbicide-tolerant crops are included in this because they do not actually produce a pesticide, which would be the responsibility of the EPA (discussed below). Instead, herbicide-tolerant crops are engineered to produce a protein which are resistant to a specific herbicide sprayed on them. An example of this is glyphosate-tolerant soybean, corn, and canola. The USDA is comprised of several agencies that are all designed to advise different groups to meet their specific needs. An example is the Animal and Plant Health Inspection Agency (APHIS). The Animal and Plant Health Inspection Agency regulates the movement, importation, and field testing of genetically engineered organisms through permitting and notification procedures. APHIS also is responsible for inspecting veterinary biologics production. A few other agencies included within the USDA are the Forest Service, Risk Management Agency, and Agricultural Research Service. The EPA's mission is to protect the quality and health of the environment. In terms of protecting the environment from biotechnology products, their main focus is on the regulation of pest management products. There are three categories of pesticides the EPA reviews for registration. The three pesticide categories are antimicrobials, biopesticides, and conventional pesticides. Before a pesticide can be used, it has to be approved by the EPA to ensure that is it not harmful to humans or the environment. For this article, we will focus on the EPA's role in regulating biopesticides, specifically Plant-Incorporated Protectants (PIPs). Biopesticides consist of biochemical pesticides, microbial pesticides, and PIPs. Plant-Incorporated Protectants are pesticidal substances produced by plants containing added genetic material. When regulating PIPs, the EPA regulates the new protein and its genetic material, not the plant itself.

Regulation

Biotech crops go through rigorous testing for several years before they are ever put on a farmer's field. In the U.S., there are nine steps in the regulatory process that usually take seven to ten years to complete, which is far more rigorous of a process than any conventionally derived food ever goes through. The USDA/APHIS is responsible for ensuring that any released organisms are safe and will not pose any negative effects on the environment, whereas the EPA evaluates the safety of transgenic plants containing PIPs. They consider where and how the protein is produced in the plant and its potential to be toxic to humans and other organisms. Before pesticides can be marketed, the EPA considers factors regarding PIPs such as:

- Risks to humans
- Risks to non-target organisms and the environment
- Gene flow potential
- Insect resistance management plans

The EPA also requires certain tests to be conducted to identify any risks to humans and the environment. These tests are listed below.

- Identification of new genetic material and all new proteins
- Mammalian toxicity testing of all new proteins
- Comparison of new proteins to known toxins and allergens

- Toxicity testing on birds, fish, earthworms, insects such as bees, ladybird beetles, and lacewings
- Toxicity testing on insects related to target insect pests
- Length of time required for the new proteins to degrade in the environment

The development of a transgenic plant requires researchers to meet with one or more regulatory agency (EPA, USDA/APHIS, or FDA) to discuss the outline of the projects and decide which studies will be necessary to ensure safety of the end product. Throughout the study the developer sends the data of the ongoing research to the appropriate agency. For many biotech crops, the product then goes through a deregulation process in which the USDA/APHIS reviews the data received from the developer and decides whether or not the transgenic plant causes a negative outcome in the environment by genetically crossing with native species or out-competing them. They also try to ensure that the transgenic plant does not have any negative effects on wildlife, and does not become a "super-weed." The overall goal of the regulatory agencies is to ensure that biotech crops are as safe as their non-genetically modified counterparts.

Preventing a "Super-weed" and Insect Resistance

A "super-weed" can occur in two ways: (1) when pollen from a genetically modified crop pollinates a weedy relative and that weedy relative then becomes resistant to the herbicide that was originally produced to kill it, and (2) when an herbicide-tolerant crop escapes cultivation and becomes a weed. For this reason, the EPA developed a regulatory framework to ensure that the appropriate tests are done to maintain that biotech crops are safe for the environment. They have not approved a genetically engineered crop that is known to have a weedy relative on the continent in which it grows. Many of our major crops, such as cotton and corn, are not native species to North America and their weedy relatives do not grow near them. For the crops that do have weedy relatives, the concern of a super-weed is not a new problem. For years plant breeders have bred plants to be superior agriculturally to any wild plant, and when grown near their weedy relatives they can already cross-pollinate and transfer their genes. An example of this is sunflowers grown for agricultural purposes can, and do, cross pollinate with wild sunflowers. A concern with insect resistant crops, such as Bt corn, is that insects feeding upon the corn will eventually become resistant to the transgenic plant, defeating the purpose of growing the insect-resistant plant.

The EPA has taken precautions to help prevent this scenario by requiring farmers who grow genetically engineered crops to also grow conventional crops nearby as "refuges" where insects aren't exposed to the genetically modified crop that contain insect resistance. In theory this allows non-resistant insects to continue breeding and remain present in the environment, and mate with any resistant survivor of the genetically modified crop. A non-resistant insect that mates with a resistant insect helps to ensure that the dominant gene of non-resistance evolves from one generation to the next, which maintains a population of insects that will still be susceptible to the insect-resistant crops. There are also issues surrounding the topic of harming beneficial insects with genetically engineered plants. To date, there have been no observed effects on beneficial insects. It is important to keep in mind that conventional methods of spraying crops for insect pests kill many beneficial insects and are not as specific on insect targets as transgenic crops resistant to a certain species of insect.

Helping the Environment

Over the years, as biotech crops have been grown, the notion that transgenic crops are actually improving the environment and may be safer to use has been a topic of debate. The "healthy environment" idea has arisen because the amount of herbicides put on herbicide-resistant crops is usually less than conventional crops and the herbicides used (mainly glyphosate) are often less toxic than other chemicals used on agricultural crops. A study conducted by Gianessi et al. (2002), confirmed that soybeans, corn, cotton, papaya, squash and canola, all crops using genetic engineering that are currently on the market, helped to reduce the pesticide volume by approximately 46 million pounds during 2001. Genetically engineered crops have an awesome potential to benefit people from all parts of the world. For this technology to be successful, regulatory processes are important to establish and implement. As biotechnology evolves, the regulatory agencies will surely have to re-structure their process to best ensure that products are safe and to be mindful so as not to inhibit the technology from moving forward. For the years that these biotech crops have been on the market, it is uncertain even to the people who have performed all the safety checks, what the long term results these plants will have on the environment, and humans. All new ideas or technologies will have risks associated with them, but the technologies that last are those whose benefits outweigh the risks. To date there have been no confirmed adverse health effects from biotech products. The regulatory agencies involved are doing what they believe is best at the moment to ensure that each product is safe.

Questions abound regarding the safety of genetically manipulated agricultural products. These questions and concerns are usually focused on laboratory or "genetically engineered" products and not those developed through traditional plant breeding technologies involving the manipulation of traits using hand pollination. While selective breeding and the production of food by microorganisms have been around for thousands of years, some people may find it mysterious or worrisome to move genetic material from one cell into another cell that may also be derived from a different species. Some of their questions range from concerns about allergic reactions, protecting the environment, the transfer of antibiotic resistance from one species to another, to fears about reducing the amount of biodiversity in the world.

The Environmental Protection Agency (EPA), the US Department of Agriculture (USDA) and the Federal Drug Adminisitration (FDA) all have regulations that impact agricultural biotechnology. Which agency or agencies are involved depends on the type of technology used. An introduction of new genetic material into a different species will likely involve additional restrictions and pre-market testing than does the use of more traditional methods such as selective breeding. Most of these regulations relate to genetically engineered materials where genes from one cell have been inserted to create a new variety. Field testing for new varieties produced by selective pollination are not usually required to demonstrate consumer safety. The U.S. Department of Agriculture (USDA) and the Environmental Protection Agency (EPA) have the roles of imposing safety regulations and/or performance standards for pesticides, herbicides and genetically engineered crops. An example is the insertion of genetic material into plants to convey resistant to the herbicide, Roundup®(Monsanto). Roundup® can be sprayed on fields to eliminate weeds and other unwanted plants while not affecting the desired crop.

FOOD BIOTECHNOLOGY IN THE UNITED STATES: SCIENCE, REGULATION, AND ISSUES

Genetic modification of agricultural crops promises the availability of food products with more desirable traits, such as higher quantities of vitamins or lowered amounts of saturated fats for consumers, reduced use of pesticides and other chemicals for environmentalists, and increased yields for growers. Traditional plant breeding, the conventional method to modify plants' genes, has produced similar benefits. But recent biotechnological innovations allow scientists to select specific genes from one plant or animal and introduce them into another to confer desirable traits. This produces the new plant or animal more quickly than conventional methods, and creates plants and animals with traits not found previously in nature.

Proponents argue that advances in genetics and new technologies can produce foods with greater yields to feed the growing world population in the 21st century. Critics are concerned that this technology produces uncertainties about potential long-term impacts on public health and the environment, and increases problems related to trade. The federal government under statute and through regulation attempts to ensure that food manufacturers produce safe products. However, some Members of Congress have asked whether federal regulations adequately manage genetic engineering risks to public health and safety, and the environment. Nine bills introduced into the 106th Congress asked that the Administration pay more attention to these risks. Similar bills are likely to be introduced into the 1067th Congress which will call for the same additional scrutiny and research.

This report discusses the science of food biotechnology, and the federal structure by which it is regulated. Because

U.S. farmers are adopting this technology at a rapid rate, some observers advocate a more active role for the federal government to ensure that farmers have equal access to this technology. Others believe that federal officials should play a more active role in protecting the environment, funding more research, and participating in international trade negotiations to ensure that trade continues to expand for genetically engineered crops. Trading partners often label food products that have been genetically modified as genetically modified organisms (GMOs). Many of those partners have labeling requirements for GMOs to allow consumers the "right to know" their food content.

Several congressional committees oversee federal governance of genetically engineered foods and biotechnology. In the Senate, food biotechnology issues are considered by the Committees on Agriculture, Nutrition, and Forestry; Health, Education, Labor and Pensions; Environment and Public Works; and Governmental Affairs. In the House, food biotechnology issues are considered by the Committees on Agriculture; Energy and Commerce; Government Reform; and Science. The Appropriations Committees of both the House and Senate have oversight responsibility on how the major federal agencies set and enforce policies affecting the safety of genetically engineered foods.

The Science of Genetic Engineering

Biotechnology is defined as the use of biological processes for the development of products such as foods, enzymes, drugs, and vaccines. Biotechnology is the new label for a process that humans have used for thousands of years to ferment foods such as beer, wine, bread, and cheese. In these cases, biological processes are used to alter raw food products to produce more stable foods. Presently, the term biotechnology is used to describe genetically engineered foods that contain genes modified by modern technologies.

When plants breed in the wild, genetic changes occur spontaneously and result in a haphazard transfer of a large number of genes within closely related species. Traditional plant breeding is more selective and creates plants with improved yields or some other desirable trait among closely related species. Often, unwanted genes conferring undesired characteristics may be transferred along with the desired characteristics. Breeding itself takes time due to the need to backcross each plant to eliminate undesirable traits. Modern genetic engineering gives greater control of the process and transfers specific genetic material into the cells of a plant. This method can reduce the likelihood of unexpected results. Also, plant breeders can use the newer genetic techniques to move genes among unrelated species to yield plants with novel traits that could not be produced by traditional breeding.

Several techniques are employed by genetic engineers. All involve DNA transfers from one plant or animal to another. DNA, deoxyribonucleic acid, is the chemical from which genes are constructed. Specialized laboratory techniques, generally referred to as recombinant DNA (rDNA) techniques are used to manipulate DNA isolated from animal, plant, or microbial cells and to introduce the engineered DNA sequences into another organism. The laboratory techniques of genetic engineering may involve direct cellular uptake of DNA, forced introduction of DNA into a cell, or the use of a non-pathogenic carrier to transmit genetic material into a cell.

Additionally, some microbial cells in immediate contact can transfer DNA directly from one cell to another. Plants may also be genetically modified by fusion of whole cells. After plant cells are genetically modified, tissue culture techniques are used to encourage growth of the modified cells into whole plant systems with leaves, stems, and roots. New food plant characteristics depend on which genes are transferred, whether these genes are switched on (expressed), and the interaction between genes and the cellular environment in which they reside.

Use of Biotechnology to Produce Food

The first wave of agricultural biotechnology food products is not substantially different from those foods already familiar and available to consumers. These modified agricultural commodities have, for the most part, directly benefitted agricultural producers with increased yields and reduced production costs. According to an industry trade association, genetically engineered food crops planted and marketed by U.S. farmers include corn, canola, rice, tomatoes, potatoes, and soybeans. Peppers, sunflowers and peanuts are in the pipeline for approval. Other genetically engineered food crops, such as sugar beets, wheat, squash, papayas, berries, bananas, and pineapples, have been developed in laboratories, and will go through the approval process for marketing within the next few years. Non-food plants that are being genetically modified include trees, for pulp wood, and cotton, although cotton seed oil may be used in food products.

Genetically Modified Whole Food Products. Some of the products referred to above have been genetically modified to be either resistant to pesticides, or to make their own pesticides. Those modified to resist pesticides allow farmers to use a herbicide for weed control without killing the food crop. This use reduces competition from weeds for nutrients and increases yields. Other products of bioengineering produce insecticides within their cells. Crops such as corn, cotton, and potatoes, have been genetically engineered to make their own pesticide. Farmers rapidly accepted these genetically engineered field crops with increasing acreage of herbicide resistant crops. The use of these plants reduces the need to plow, decreases the amount of chemical herbicide needed, produces higher yields, and can deliver a cleaner and higher grade of grain and product. Of the total 1998 crop, approximately 25% of planted corn was genetically modified, 38% of planted soybeans, 45% of cotton, and 42% of canola. In 1999 the percentage of acres planted with corn grew to 37% but declined to 25% in 2000. In 1998 in the United States, herbicide tolerant soybeans became the dominant bioengineered crop, with 4.1 million acres planted (about 36% of the total U.S. acreage planted in soybeans). In 1999, bioengineered soybean plantings increased to 47% and continued to increase to 54% in 2000. Cotton has followed a similar trend. Herbicide resistant crops of soybeans, corn, and cotton accounted for most of the acreage planted in biotech crops.

Various companies have targeted different genetic modifications to produce tomatoes that remain firm for a longer time and are reputedly more flavorful than traditional tomatoes. Calgene's Flavr Savr tomato was one of the first genetically engineered consumer-ready foods to receive federal approval for production and marketing in the United States. Another tomato engineered by Zeneca, used for production of a reduced-price tomato paste for sale in the United Kingdom, was withdrawn due to consumer protests.

Biotechnology is also used to produce experimental "transgenic" animals, in which the genetic material has been deliberately modified and to produce "clones" in which animals are reproduced artificially but the DNA is not modified. In agriculture, transgenic animals may be altered to produce higher yields of specific products (meat, milk, etc.) or to bring about commodities with enhanced characteristics, such as less cholesterol or reduced fat content. Although cloning has been used to reproduce animals for scientific purposes since the 1950s, its usefulness for the reproduction of identical livestock animals was only recently investigated. In 1995, sheep were cloned from embryonic cells in Scotland. In 1996, a substantial breakthrough followed when a sheep, Dolly, was cloned from an adult, nonembryonic cell. Japanese scientists are creating high-value beef cattle through cloning. They have successfully cloned at least 19 calves from adult bovine cells. Because the cost of some premium beef roasts can be between $100 and $200 per pound in Japan, the Japanese cattle industry can support the expense of cloning prize beef cattle. But even with those prices, the cost of genetically engineering cattle on a large scale could be prohibitive.

At this time, most of the research on transgenic animals in agriculture is experimental. The Biotechnology Industry Organization (BIO), an industry trade organization, estimates that the only transgenic animals that will be marketed within the next 6 years are transgenic fish that can grow to market size more rapidly than traditional farm-raised fish. Experiments are also being conducted to produce transgenic animals that yield human pharmaceuticals such as vaccines, growth hormones, blood-clotting factors, monoclonal antibodies and other drugs. Enhancement of milk from cows, goats and sheep to contain these drugs is the target of much of this research. Animal products also have been changed by biotechnology. A food processing agent, chymosin (also called rennin) is an enzyme required to manufacture cheese. It was the first genetically engineered food additive to be used commercially. Traditionally, processors obtained chymosin from rennet, a preparation derived from the fourth stomach of milk-fed calves. Scientists engineered a non-pathogenic strain (K-12) of *E. coli* bacteria for large-scale laboratory production of the enzyme. This microbiologically produced recombinant enzyme, identical structurally to the calf derived enzyme, costs less and is produced in abundant quantities. Today about 60% of U.S. hard cheese is made with genetically engineered chymosin. In 1990, FDA granted chymosin "generally-recognized-as-safe" (GRAS) status based on data showing that the enzyme was safe. The final enzyme product is purified by removing potentially harmful substances, including the gene for antibiotic resistance used to engineer the microorganism that produces chymosin.

Bovine somatotropin (BST), also known as bovine growth hormone (BGH), occurs naturally in cows. When recombinant bovine somatotropin (rBST), is injected supplementally into dairy cattle, milk production may increase 10% to 15% (see text box). Genetically engineered microorganisms produce a consistent and affordable supply of this hormone as opposed to isolating the compound from limited bovine sources. According to an industry trade association, it is possible that up to 30% of U.S. dairy cows are injected with recombinant BST to increase milk yield. **Future Products.** Experts indicate that the "second wave" of genetically modified food products will target consumer and animal health issues and improve the nutrition content of certain foods. For example, vitamin A shortages that are a significant health concern in developing countries could be addressed by increasing the vitamin A content in canola oil. Genetically engineered soybeans, peanuts, and sunflowers may contain reduced levels of saturated fats or have altered fatty acid compositions for enhanced health benefits and to improve vegetable oil stability without the need for chemical hydrogenation. (Hydrogenation is used to create fats that are solid at room temperature — margarine, solid Crisco, etc.) Fruits may also be genetically engineered to contain vaccines. These fruits could be delivered without special care to developing countries which often don't have refrigeration necessary for current vaccines. Genetically engineered salmon, trout, and flounder achieve market size in half the time of non-genetically engineered fish. Commercialization of such fish may help curtail over-fishing of native fish, as well as reduce consumer prices. In addition, scientists are using genetic engineering to make high-value products, such as special oils and chemicals. Rapeseed plants have been genetically modified to produce 35% more of a fatty acid, laurate, for use in soaps, detergents and other household items. Other plants and animals are being engineered to produce pharmaceuticals, specialty chemicals, and biologic agents.

Federal Responsibilities for Regulating Genetically Modified Foods

During the 1970s, the development of new techniques for transferring genes raised concerns about potential hazards. At the Asilomar Conference in February 1975, scientists working with this technology tried to reach a consensus to selfregulate research involving rDNA technology until its safety could be assured. The National Institutes of Health (NIH) became involved in 1976 when it published research guidelines using rDNA techniques. Until 1984, the NIH Recombinant DNA Advisory Committee was the primary federal entity that reviewed and monitored DNA research. However, a legal challenge forced the Reagan Administration to consider and propose policies to guide activities of federal agencies responsible for reviewing biotechnology research and its products. In 1984, the White House Office of Science and Technology Policy (OSTP) published the "Coordinated Framework for Regulation of Biotechnology," a framework proposing that genetically engineered products would continue to be regulated according to their characteristics and novel features and not by their method of production. It also proposed that new biotechnology products be regulated under the existing web of federal statutory authority and regulation. In 1986, OSTP finalized this framework. The framework identified lead agencies to coordinate activities when and if jurisdictions overlapped. For example, the Food and Drug Administration (FDA) is responsible for regulating food and feeds in the market that have been modified through genetic engineering. The U.S. Department of Agriculture (USDA), Animal and Plant Health Inspection Service (APHIS), regulates importation, interstate movement, and environmental release of transgenic plants that contain plant pest components. It licenses, through permits, the field testing of food crops prior to commercial release. But agencies' responsibilities overlap as some plants have been modified to contain plant-pesticides. The Environmental Protection Agency (EPA) registers certain pesticides produced in transgenic plants prior to their distribution and sale and establishes pesticide tolerances for residues in foods. APHIS and EPA together established that APHIS would regulate the plant itself, particularly those plants engineered to resist herbicides, and EPA would regulate the pesticide used on the herbicide resistant plant. They also together established procedures to review and approve field tests of modified plants and microorganisms. FDA has post-market authority to remove a food from the market.

Food and Drug Administration

The Federal Food, Drug, and Cosmetic Act (FFDCA) gives FDA broad authority to regulate foods by prohibiting the entry into interstate commerce of adulterated or misbranded foods. It is the legal responsibility of food manufacturers to produce foods that are not adulterated, unsafe, filthy, or produced under unsanitary conditions. FDA has authority to inspect foods and food facilities, both domestic and imported, to ensure that they are manufactured and held under acceptable conditions and are properly labeled.

FDA can seize products or request that they be removed from the market if they do not meet federal requirements.

The Act also requires that "food additives" not be marketed unless they have received approval from FDA. But substances added to foods that are considered generally-recognized-as-safe or GRAS substances do not need agency approval. First articulated in the OSTP framework document in 1986, FDA determined in 1992 that bioengineered foods pose the kinds of scientific and regulatory issues that are not substantively different from those raised by non-bioengineered foods. Thus, FDA regulates foods that have been genetically modified or engineered no differently than foods created by conventional means.

In a May 1992 policy statement FDA described how its regulatory authority applies to new plant varieties and derived food products, including those developed through genetic engineering. The agency decided that companies developing genetically engineered foods would have to go through a special review in FDA only if : ! the gene transfer produces unexpected genetic effects; ! the levels of toxicants in the food are significantly higher than present in other edible varieties of the same species; ! nutrients in the bioengineered food differ from those in traditional varieties; ! the sources of the newly introduced genetic material come from a food plant associated with allergies; ! the food from the new variety differs significantly in composition from food of comparable varieties; ! the food contains marker genes that theoretically may reduce the therapeutic effects of clinically useful antibiotics; ! the plants are developed to make substances like pharmaceuticals or polymers, and will also be used for food; or ! the food to be used for animal feed has changes in nutrients or toxicants.

So far, most genetically modified foods have not required

pre-market approval, although a few have had their composition changed significantly such that they were labeled differently. Most proteins from genes transferred into foods to give them new traits are either GRAS or otherwise exempt from regulation. A GRAS substance is excluded from the definition of a food additive. FDA has, however, instituted a voluntary consultation process whereby the developer can resolve any safety or regulatory issues prior to marketing. The agency has proposed keeping this process voluntary when it published a proposed rule in January 2001. The proposed rule would require that a food company notify the agency 120 days prior to marketing a bioengineered food and supply the agency with safety test data. FDA strongly urges companies to consult the agency prior to the mandated notification deadline in order to ensure agreement on the types of safety testing that will be needed. By the end of 2000, developers had consulted with FDA officials on 44 products: 12 times for corn; 5 for tomatoes; 8 for canola; 5 for cotton seed; 4 for potatoes; 3 for soybeans; 2 each for sugar beets and squash; and once for flax, radicchio, and papaya.

Case Study of a Commercial Genetically Modified Food: Flavr Savr Tomato

Although a few corporations have made substantial, long-term investments in the application of biotechnology to foods, not all genetically modified foods are commercially successful. Calgene's Flavr Savr tomato, for example, had limited technical success and did not meet commercial expectations. Calgene genetically modified a strain of tomato to reduce activity of a particular enzyme (polygalacturonase) that affects softening of outer tissue during ripening. Because the genetically modified tomato had less of this enzyme, it could remain longer on the vine prior to harvest thereby enhancing its tomato flavor. However, Calgene chose to manipulate genetically a tomato strain that had qualities more useful for processing than for the fresh market.

The Flavr Savr never achieved commercial success because it cost more and did not taste better than competing cheaper tomatoes.

After its founding in 1980, Calgene conducted basic molecular biology research to investigate optimum techniques for the genetic manipulation of foods. In the mid 1980s, Campbell Soup co. decided to support Calgene's research efforts to produce a genetically modified tomato, presumably for use in Campbell's products.

In 1989, officials at Calgene initiated discussions with FDA regarding the regulatory status of the tomato. FDA responded in 1992 by publishing its policy on foods derived from transgenic plants. Essentially, FDA decided to regulate genetically modified foods as any other food derived from traditional practices. The genetic technology needed to alter the tomato used an antibiotic marker that produced very small amounts of a non-tomato protein in Flavr Savr. According to FDA, the protein was viewed as a food additive since it changed the tomato's composition. Calgene officially petitioned FDA in January 1993 to allow the presence of this protein as a food additive. In May 1994, FDA approved Calgene's petition. This approval opened the way for commercial marketing of the Flavr Savr tomato. However, in 1993 with significant public opposition to the genetically engineered tomato Campbell Soup Co. decided not to use genetically modified tomatoes in its products. Calgene then began efforts to market Flavr Savr as a freshmarket tomato rather than for use in processing. However, the tomato bruised easily and was less firm than expected.

This characteristic caused production, transportation, and distribution problems. Competition from new tomato strains bred by traditional methods was an additional obstacle. As reports of problems with commercialization of Flavr Savr grew, Calgene's financial condition weakened. In June 1995, Monsanto acquired a 49.9% equity stake in Calgene through the purchase of Calgene stock. In August 1996, Monsanto acquired controlling interest in Calgene. Monsanto emphasized other research programs at Calgene and subsequently moved control of Flavr Savr to another of its subsidiaries, Gargiulo Inc. in Naples, FL. According to industry sources, Gargiulo discontinued the effort to commercialize Flavr Savr.

In contrast, a British company, Zeneca, succeeded in marketing a genetically modified tomato in England. Now grown in California and processed into tomato paste for sale in the United Kingdom (U.K.), the Zeneca tomato has a label declaring that the paste is produced from genetically modified tomatoes. This product has since been withdrawn from the market because of consumer protests.

Development of this tomato used genetic technology virtually identical to that used by Calgene. Zeneca chose to genetically modify similar ripening-related enzymes in a tomato strain that had desirable processing characteristics. The Zeneca tomato yields a paste that is perceived as thicker and more flavorful than other tomato pastes. When combined with a comparatively low price plus marketing efforts through established market chains in England, this product has become the best selling tomato paste in the U.K.

Source: J.H. Maryanski. "FDA's Policy For Foods Developed By Biotechnology." *Genetically Modified Foods: Safety Issues* (Ed.s, Engel, Takeoka and Teranishi). American Chemical Society, Symposium Series 605, Ch. 2, 1995. p. 12-22; "Biotechnology of food." *FDA Backgrounder*. May 1994.

On a food product's label, the FFDCA requires producers of foods to describe the product by its common name and reveal important facts associated with claims made or suggested. The agency interprets the FFDCA as not giving it authority to mandate labeling based solely on a consumer's "right to know" the method of production if the final product is considered "safe." Therefore, the agency does not mandate labeling to indicate the method by which a new variety was developed (e.g., that it was genetically engineered). However, the FFDCA does require that all information on labels be truthful and not misleading. Special labeling may be required if the developed food significantly differs from its conventional counterpart such that the common name would no longer apply. For example, FDA required the renaming of a soybean oil whose fatty acid composition had been altered by engineering. The new name, "high oleic acid soybeans," describes what is different about the oil but not its production method.

FDA requires special labeling for foods if they pose special safety or usage issues. For example, if a food had a new protein introduced into it to which people were allergic, FDA would require the label to reveal that information. In its 1992 policy statement, the agency noted that labeling would be required if genes were introduced from foods that were commonly allergenic, unless the developer could scientifically demonstrate that the protein was not responsible for the allergenicity of the original food. Examples of commonly allergenic food include milk, eggs, wheat, shellfish, tree nuts, and legumes. In one case, a developer demonstrated that DNA transferred to soybeans from a Brazil nut caused the production of a protein responsible for an allergic reaction. Consequently, the private developer discontinued research on that particular modified soybean. FDA has asked that developers of genetically modified food demonstrate that the introduced proteins do not share structural similarity to known allergens and are not resistant to digestive enzymes and acid.

On January 18, 2001, FDA published in the *Federal Register* a "Draft Guidance for Industry: Voluntary Labeling Indicating Whether Foods Have or Have Not Been Developed Using Bioengineering" and is seeking comments. In this document, FDA reaffirmed that it believes most genetically engineered foods are substantially equivalent to their conventional counterparts, and it decided it would not require special labeling of all bioengineered foods because it believes that the use of bioengineering, or its absence, does not itself cause a material difference in the food.

However, the agency did suggest that because of the strongly divergent views on labeling, manufacturers may consider providing more information about bioengineered food. The information given, however, must be shown to be truthful and not misleading. To avoid false or misleading statements about the absence of bioengineered ingredients (because there are no established threshold levels of bioengineered constituents or ingredients in foods), or to avoid implying that one food is superior to others, FDA suggests not using statements such as "GM free" or "biotech free." The agency does suggest the word "biotechnology" is preferred by some consumers over "genetic engineering" or "genetic modification." It also claims that if validated testing is available, it can be used to verify whether the label is truthful. Or manufacturers could keep records to document the reasons why a food's label is truthful. Comments on the draft guidance are requested.

That same day, January 18, 2001, FDA proposed to make mandatory a notification system whereby a food company would notify the agency 120 days prior to releasing a genetically engineered food onto the market. Along with the notification must be data that show that the is safe. FDA is not proposing to require that genetically engineered foods undergo a "food additive" review — the extensive pre-market approval process. The agency is encouraging industry to consult FDA about the types of tests that could be needed to prove safety. FDA would then have time to review the company supplied data and would make it available to the public for comment. Proprietary information would not however be released to the public.

Critics of this proposed rule think that independent

testing of the safety data is necessary to restore consumer confidence in this food. They assert that the testing should be done either by the agency or independent contractors but not by the company that has a vested interest in showing only the product's safety.

U.S. Department of Agriculture (USDA), Animal and Plant Health Inspection Service (APHIS)

APHIS issues permits for the importation and domestic interstate shipment of certain plants, animals, and microbes that have the potential for creating pest problems in domestic agriculture. The agency has historically regulated pests that attack plants — any living stage of any insect, mite, nematode, slug, snail, protozoa, and/or other invertebrate animal, bacteria, fungi, or parasitic plant or reproductive part. It is also interested in viruses — infectious substances that would directly or indirectly injure or cause disease or damage plants or plant parts or any processed, manufactured, or other plant products. For new plants that could become pests, APHIS issues site specific permits for field tests or for release into the environment. The agency reviews permit applications and prepares an environmental assessment in which it evaluates the probable environmental impact of the release. The permit application process requires that the developer disclose information about the development of the plant and that appropriate facilities and control measures are in place during transport and field tests.

If the agency reaches a "Finding of No Significant Impact" (FONSI), a permit is issued. Before decisions are made, APHIS seeks concurrence with states on regulatory actions.

In 1993, APHIS introduced an expedited procedure for approving limited permits so that field testing of six crops could begin without a completed formal application that included an environmental assessment. For genetically engineered plants that meet certain eligibility requirements and performance standards, the sponsoring company need only submit a "notification" letter to the agency, a modified and abbreviated application which describes the gene, where the tests will take place, and the characteristics of the plant. The agency has 30 days to process the application before the sponsor can proceed with the field test. In 1997, APHIS expanded the expedited procedure to cover many more crops; by 1998, 99% of all applicants used the expedited process.

After tests are completed and an application is submitted, APHIS has 120 days to decide whether the product poses a risk of being a plant pest and whether a product is ready for full "release" onto the market. APHIS then completes an environmental assessment before making its decision.

Case Study of a Genetically Modified Hormone: rBST

Bovine somatotropin (BST), also referred to as bovine growth hormone (BGH), is produced within the pituitary gland of all cows. It helps in the lactation process and is a normal trace constituent of milk.

Farmers inject doses of the genetically modified hormone (called commercially Posilac) into cows to enhance milk yields and lengthen the lactation cycle. The result is an increase in milk yields of up to 15%. Prior to the 1980s, BST treatments were experimental and costly, since extractions of bovine tissues were the only source of the compound. With recombinant DNA technology, the supply of recombinant BST (rBST or rBGH) is more abundant and less expensive. Critics are concerned that excessive amounts of rBST could compromise human and animal safety. They state that FDA did not provide adequate review of data to establish safety of the product prior to its approval. Six scientists within Health Canada, the Canadian federal agency involved in approving drugs, contend that the drug may not be safe for human consumption. However, the expert panel charged by Health Canada to review the drug found no risks to human safety. On January 14, 1999, Health Canada rejected approval of rBST based on concerns for animal health.

In June 1992, a joint expert committee of FAO/WHO concluded that rBST is safe for use and that Maximum Residue Limits (MRLs) are unnecessary. After a second review in February 1998, the committee arrived at the same conclusion. On March 10, 1999, the EU Scientific Committee on *Animal Health and Animal Welfare* recommended that the current EU ban on the use of rBST should continue in effect.

In 1989, Monsanto petitioned FDA's Center for Veterinary Medicine to approve an rBST product as a new animal drug. The review process took 4 years and was more extensive than most approval processes due to the controversial nature of the product. In August 1990, FDA published a review of data on rBST and concluded that it "presents no increased health risk." An August 1992 GAO report suggested that there may be an increase in mastitis in cows treated with rBST. The report also suggested there could possibly be indirect human health effects from residues of antibiotics used to treat cows for mastitis (udder infections). On March 31, 1993, an advisory committee of FDA's Center for Veterinary Medicine concluded that adequate safeguards are in place to prevent unsafe levels of antibiotic residues from entering the milk supply due to increased mastitis in rBST treated cows.

FDA approved the rBST product, Posilac, on November 5, 1993, with the stipulation that its developer, Monsanto, conduct a post approval monitoring program (PAMP) to provide further information related to possible effects on animal health experienced by rBST treated cows. FDA publishes occasional PAMP updates to summarize clinical manifestations associated with rBST treated cows. Since 1994, there have been 1,235 reports of adverse reactions in cows treated with Posilac although FDA states "the reported clinical manifestations are known to occur in dairy cattle not supplemented with Posilac." It also indicates that the number and types of reported effects raise no new animal health concerns.

On February 7, 1994, FDA offered interim guidance on labeling of milk from untreated cows, since some companies wanted to label their milk products as "BST-free." Products may be labeled as coming from animals not treated with rBST, but since BST is a normal constituent of milk, FDA determined that it is misleading to label milk as "BST-free." In May 1994, FDA's Food Advisory Committee and Veterinary Medicine Advisory Committee discussed whether foods derived from cows given supplemental rBST should be labeled as such. The committee report states "deliberations indicate that any method for instituting labeling for food from BST-supplemented cows would have to resolve many difficult scientific and policy questions." In 1999, FDA requires no labeling of milk products produced from cows supplemented with rBST.

On December 15, 1998, the non-profit Washington, D.C.-based Center for Food Safety petitioned FDA to withdraw approval of rBST citing possible health effects not addressed by FDA. As of January 2001 this issue is pending within the agency.

U.S. Department of Agriculture, Food Safety and Inspection Service (FSIS)

Developers of transgenic animals must submit data to

FSIS to prove that the livestock and poultry involved in biotechnology experiments are not adulterated and can be slaughtered and sold as food with other beef and poultry. Prior to approving slaughter and sale, FSIS inspectors look at the number, age, sex, and other factors. The ultimate disposition of transgenic animal carcasses, whether by rendering, slaughter for food, or other disposal is of concern to regulators.

Environmental Protection Agency

The EPA regulates pesticides. Pesticides are broadly defined as any substance or mixture of substances intended for "preventing, destroying, repelling, or mitigating" pests. EPA currently refers to "plant-pesticides," now called "plantincorporated protectants," as plants that produce pesticides within their tissues. Scientists have also genetically engineered plants that are resistant to specific herbicides. Although herbicide resistant plants are not "plant-incorporated protectants," they are subject to EPA regulation since they can affect the use of herbicides.

EPA regulates plant-incorporated protectants, under the Federal Insecticide, Fungicide and Rodenticide Act (FIFRA) and FFDCA. Under FIFRA, EPA determines the risk the pesticide substance in the plant poses to humans and the environment and approves registration of those substances for particular uses that will not generally cause unreasonable adverse effects. This determination involves balancing risks from the pesticide with benefits associated with its use. A pesticide (including a plant-incorporated protectant) cannot be sold or distributed in the United States unless it is registered with EPA.

If the plant producing the plant-incorporated protectant is a food crop, EPA must establish a "safe level" of pesticide residue allowed, a tolerance level, under the authority of

Section 408 of the FFDCA. A "safe level" of the pesticide residue is defined as that level at which there is "a reasonable certainty that no harm will result from aggregate exposure to the pesticide chemical residue, including all anticipated dietary exposures and all other exposures for which there is reliable information." Because no tests of the registered (approved) plant-incorporated protectant have shown toxicity to humans so far, EPA has given them an exemption from the requirement for a tolerance level.

The delta-endotoxins are proteins, one of the many toxins that may be naturally produced by the bacterium, *Bacillus thuringiensis,* and are species-specific, affecting only certain insects. They are also virtually harmless to humans and animals. When a susceptible insect consumes the protein, its digestion is severely disrupted, further feeding stops, and the pest eventually dies, usually within 2 days. In 1994, EPA proposed a rule to refine its regulatory oversight of plantincorporated protectant. As a part of the rule, EPA included in its definition both the plant-incorporated protectant and the genetic material transferred into the plant.

Under this rule, EPA proposed to exempt several categories of plant-incorporated protectant from FIFRA and FFDCA section 408 requirements. Under this rule, EPA proposed to exempt: (1) plant-incorporated protectants derived from plants sexually compatible with the recipient plant; (2) plant-incorporated protectants that act by primarily affecting the plant; and (3) plant-incorporated protectants based on a coat protein from a plant virus. These rules were designed to exempt certain categories of substances that EPA believes are low risk based on familiarity and presence in the food supply, e.g., plant hormones and coat proteins from plant viruses. The agency believed that these proposed exemptions would result in little or no effects on plants produced by conventional plant breeding. They could, however, affect research into innovative plants, according to critics, because of the additional regulatory burden that the rule would represent. The rule was finalized on January 18, 2001.

EPA's Final Plant-Incorporated Protectant Regulation.

In the January 18, 2001 final rule, EPA clarified three regulations governing plant-incorporated protectants. With this regulation, EPA intends to ensure that plant-incorporated protectants derived from genetically engineering will meet federal safety standards under both the FFDCA and FIFRA. EPA will also set food tolerances for residues of plant-incorporated protectants (or give exemptions) found in foods. The agency also exempted from registration plant-incorporated protectants derived from conventional breeding of sexually compatible plants and genetic material necessary for the production of plant-incorporated protectants. The final rule did not exempt plantincorporated protectants derived through genetic engineering from sexually compatible plants; plant-incorporated protectants that act primarily by affecting the plant; and plant-incorporated protectants based on viral coat proteins. Both these exemptions were questioned in the National Academy of Sciences report, published in April 2000, entitled "Genetically Modified Pest-Protected Plants: Science and Regulation." EPA has asked for comments on the report's recommendations.

In September 2000, StarLink corn containing a Cry9C protein, a protein approved only for use in animal feed, turned up in taco shells. Without approval for human use, or exemption from approval, the Cry9C protein is considered an adulterant. The current regulatory system would still not necessarily have caught this adulterant in human food, even if there had been a mandatory notification and labeling program. The incident and subsequent problems with StarLink corn being found in shipments that would have been used in food, however, has led EPA to state publically that it will probably never again allow the entry of bioengineered products onto the market that cannot be used in human food. The incident also led FDA to develop sampling and testing guidance for the industry so that testing results could be used to verify the labeling of corn with or without the Cry9C protein.

Issues

Congress' attention will likely be drawn to a number of biotechnology issues due to public concern. Issues include concerns about public health, religious issues, labeling, the environment, the economic impact of this technology, and international trade competitiveness.

Labeling for Public Health and Religious Practices Public Health Concerns

Many consumers express wariness of new "supercrops" and novelty foods, fearing that introduced genes could prove allergenic or harmful to human health. For example, if new genes inadvertently caused a plant to produce toxins at higher levels than are present naturally, there could be long-term health consequences for humans. Some consumers are worried that a gene introduced into plants to protect against pests could also cause the plant to alter its pollen, thereby affecting the health of humans prone to some sensitivities.

Some critics are dismayed that FDA is placing all the responsibility on manufacturers to generate safety data, as it does normally under its pre-market approval system, and is reviewing only the conclusions of industry-sponsored studies, rather than conducting its own tests. Critics have asked that more tests be conducted for chronic effects prior to products being placed on the market to ensure that all uncertainties regarding human health be explored. Proponents argue that additional testing of genetically engineered foods is unnecessary because all foods must meet the same federal safety standards regardless of whether they are genetically engineered.

There is a growing movement among consumer groups that advocate the labeling of all genetically modified foods (GMO foods) that were produced through the process of genetic engineering. This position reflects a policy of "consumer sovereignty" or the "right to know," which supports the disclosure of all relevant information on a label so consumers may make food choices based on their own values.

Religious Practices

Others, particularly religious groups, are concerned that foods might contain genes from animals, such as swine prohibited by some religions, and they maintain that they have a right to know if foods contain those genes. Both the kosher (Jewish) and halal (Muslim) communities have mechanisms in place to determine which products are acceptable to their adherents, and thus have not concerned themselves with secular labeling issues. However, both Orthodox Rabbis and Muslim leaders have ruled that simple gene additions that lead to one or a few new components in a species are acceptable for kosher and halal law. The Muslim community has not yet resolved whether a gene derived from swine is an exception to the above acceptance. For example, both groups have raised no objection to the use of bioengineered chymosin (rennin) in the production of cheese.

The status of more significant changes in the genetic makeup of species remains to be decided. However, cloning, in particular, raises serious ethical/moral issues for religious leaders of all faiths. That discussion has involved a much broader range of clergy within the respective communities as well as other communities without dietary laws.

Labeling

Currently, no federal agency requires foods to be labeled as genetically modified. One reason is because the agency sees these products as substantiallly equivalent to traditional food products. This assumption has been supported by the courts. Under Vermont's 1995 mandatory rBST labeling rule retailers rather than producers paid the expense of enforcement to ensure that all milk produced using rBST was so labeled. A federal appeals court overturned this requirement, and since the spring of 1997, Vermont has authorized voluntary labels for rBST-free dairy products. On January 18, 2001, FDA published a draft guidance for the industry giving examples of how the food industry could label foods that have or have not been developed using bioengineering. The guidance suggests that consumers prefer labels that explain why it was used in the food or how it is different from traditional food.

The guidance suggests that all labels be "value-neutral." It also suggests that using the terms "GMO-free" or "not genetically modified" labeling could be misleading because such a label implies there is a zero level of bioengineered material which is almost always impossible to verify. The phrase "derived from biotechnology " would explain that the claim is being made and would allow consumers a choice about whether to purchase products produced by the new technology, to make better judgements about the compliance of the food product with ethical and religious beliefs, and lessen objections to its use. Critics of this draft say that all labeling should be mandatory, not voluntary. They also claim that if the food companies are not forced to label genetically engineered foods, if there are health or safety problems, it would be more difficult to trace illnesses stemming from a particular bioengineered food.

The food industry generally opposes all biotech labeling because consumers may interpret these to be "warning labels," implying that the foods produced through biotechnology are less safe or nutritious than conventional foods. Food production interests believe that consumers, thinking that a product is different from conventional foods, may not gain the benefits from foods that have been modified genetically because they are uneasy with the technology and may not try the products. The industry supports FDA's reasoning in the draft guidance. Defenders of the agencies' policies suggest that to date genetically modified foods are similar to non-GMO foods and are easily regulated under the current policy structure. If they are substantially different, these foods must be regulated as if they were food additives and receive FDA approval before being marketed. They state that current GMOs are not a risk as no scientific evidence shows an effect on human or animal health. For instance, most plant toxins are acute toxins and are well known, and tests have been developed to detect them quickly. So defenders argue that there are no proven long-term health consequences for humans. However, few long-term studies have been completed.

On May 27, 1998, FDA was sued by a group of concerned citizens over the agency's "traditional" policy on the sale of genetically modified foods (*Alliance for Bio-Integrity v. Shalala*, D.D.C., No. 1:98CV01300, May 27, 1998). The group claimed that FDA's refusal to require labeling and safety testing raises health and environmental concerns and makes it difficult to comply with religious dietary laws.

The suit identified 36 genetically modified foods being

consumed daily without the knowledge of U.S. consumers. The suit cited both the First Amendment's protection of religious freedom and the Religious Freedom Restoration Act of 1993, which requires that federal laws or regulations not impede the free exercise of religion. The plaintiffs said that FDA's policy failed to abide by the public notice and comment procedures of the Administrative Procedures Act, and allowed genetically modified foods into the marketplace without being identified as such. The suit also claimed that FDA's policy is a burden to consumers' abilities to follow religious dietary laws.

The lawsuit challenged FDA's policy that genetically modified foods are considered safe unless they contain substances identified in the 1992 policy which are allergens or would change the character of the food. The plaintiffs want the agency to carry out the same testing and safety evaluations conducted for food additive petition approval because, they argue, changes that might occur as a result of genetic engineering might include unwanted, unpredictable new toxins and/or carcinogens, elevated levels of inherent toxins and/or carcinogens, and/or degradation of nutritional quality. In particular, they wanted FDA to require the labeling of these foods because the foods have been changed "materially" and allegedly violate the FFDCA. The suit stated that "FDA is permitting unpredictable changes to the characteristics of certain foods which may be difficult for consumers to detect." On August 7, 1998, FDA, through the Department of Justice, asked the U.S. District Court for the District of Columbia to dismiss this case because it felt that such charges are not unique to genetic engineering and the same changes occur in foods that have not been modified. On September 29, 2000, the Court accepted FDA's view that genetically engineered foods did not need special labeling because of consumer demand or because of the process used to develop these foods. Nor did the court require premarket approval of the food.

Environmental Issues

Proponents of bioengineered crops claim that genetic modification can be less harsh on the environment than other technologies. They suggest that fewer agricultural chemicals might be needed to grow pesticide tolerant or insect resistant crops and that land would need less repeated tilling, which could lead to less erosion and soil infertility. Supporters of genetic modification think that new developments contribute to environmentally sustainable development and greater food production. While they believe government regulatory efforts adequately ensure consumer and environmental health, they are aware of growing consumer concerns and some want to increase transparency of the regulatory and development process.

Critics have expressed strong concerns about the long-term risks and consequences of cross-pollination and of the disruption to the "cellular ecology" of plants. They state that the U.S. policy is based on the assumption of safety but there is little research on ecological or food safety risks. Scientists have shown that genetically modified rapeseed (canola) pollen was spread to wild radish weed relatives in nearby fields. The experiment demonstrated that it was possible to create new strains of weeds resistant to herbicides. If such weeds emerged widely, farmers would need new, different, or stronger herbicides to counter their spread. Such super weeds could severely decrease crop productivity. Furthermore, some scientists have expressed concern that the widespread use of genetically engineered plants could alter the ecology of natural plant communities and of wildlife food chains. Certain seed and herbicide companies agree with these critics, their point of view shaped by the possibility of development of "super weeds," rendering their products useless.

Bt Resistance and Intervention Strategies

Concerns revolve around plants engineered to produce within their cells an insecticide called Bt that is produced naturally by strains of the bacterium *Bacillus thuringiensis*. The release of a study that showed in a laboratory that Bt corn pollen when eaten by Monarch butterfly larvae kill or stunt their growth engendered public concern. Another concern is that large scale planting of crops containing Bt might lead to faster resistance development by insects. Critics of this technology state that large-scale production of engineered corn, soybeans and other foods will cause pests to develop resistance to Bt, thereby limiting its usefulness. It is not unusual for insects to develop resistance to pesticides that have been used for long periods of time. Organic farmers, in particular, are concerned because they do not have as many crop protection tools available as conventional farmers, and the loss of effectiveness of Bt could be a serious blow to their production. Due to these concerns, on February 18, 1999, Greenpeace, the Center for Food Safety, and some organic farmers sued EPA over its registrations of plant-incorporated protectant. The suit calls on EPA to cancel all existing registrations for Bt crops, to cease the approval process for any new registrations, and to perform an environmental impact assessment analyzing the cumulative impacts from the current registrations.

Prior to these complaints, EPA determined that resistance to Bt can be slowed by requiring farmers to plant significant numbers of non-Bt seeds near the genetically modified resistant plants in designated separate areas called "refuges." The non-Bt plants allow pests to grow that will not develop Bt resistance. Since these pests will interbreed with their counterparts that eat Bt plants, there is less likelihood that a resistant super-pest could develop. If the "refuge" system is used by farmers, EPA contains a statement, "We require every grower who decides to plant YieldGard corn to sign an agreement to establish a refuge adjacent to his or her YieldGard acreage to ensure a population of susceptible corn borers to mate with any naturally resistant borers that survive exposure to the Bt." estimates that the eventual development of resistance to Bt by insect pests could be extended 10 or more years. EPA now requires the use of insect resistance management (IRM) plans for some crops. EPA originally granted all the plant-incorporated protectants conditional registrations. With the registrations for corn and cotton to expire in 2001, as a condition of re-registration, companies will be required to develop effective IRM plans for these crops. Most re-registration applications have been submitted, and EPA is evaluating the information and could impose new conditions. Critics of the technology want mandatory IRM plans, for they claim that if the plans are voluntary, farmers will either not plant the nonresistant refuges or plant refuges that are too small to be effective. There is a voluntary IRM plan in place for potatoes, as recommended by a 1995 scientific advisory panel, and Monsanto is requiring compliance by growers who contract with it for the seeds. Ongoing evaluation of IRM may result in additional mandatory requirements.

There is agreement, however, that toxic anti-pest substances produced by plants used for food must be proven safe. EPA points to an example of potato leaves that naturally contain a pesticidal substance that could cause birth defects. Since humans do not eat potato leaves, there is no need to regulate the substance; however, if spinach were genetically modified to contain the potato-leaf toxin, regulation would be needed.

Liability

There is also concern about liability. Who would pay if other crops or fields were ruined because of cross-pollination with these new seeds? An opponent of biotechnology, Jeremy Rifkin, says, "The insurance industry has quietly let it be known that it would not insure the release of genetically engineered organisms into the environment against the possibility of widespread environmental damage, because the industry lacks a risk-assessment science — a predictive ecology — with which to judge the risk of any given introduction." According to Dr. L. Val Giddings, Vice President for Food and Agriculture, Biotechnology Industry Organization, the industry representatives directly dispute this claim, and say there is insurance available.

Economic Concerns

Critics are concerned about the growing presence of large agricultural conglomerate companies controlling the supply of seeds containing bioengineered traits. Growers' profits from bioengineered crops more than tripled from 1996 to 1997. In 1996, U.S. farmers had planted mostly Bt cotton, Bt corn, and herbicide tolerant soybeans. In 1997, they expanded acreage for those crops and added herbicide tolerant cotton and potatoes. Planted acreage has continue to increase for all crops except corn. Adoption rates vary by year, crop product, and location. They depend on the level of infestation of a targeted pest. They also depend on world commodity prices and the kind of corn that importers are willing to accept.

Supporters of this technology cite convenience and lower costs as the main reasons for high farmer adoption rates. But critics complain about the consolidation of patent ownership of this new technology which they believe constitutes monopolistic power in the marketplace. Their distrust is based in part on concern about the dependency such concentration brings to the agricultural sector and the possible abuse from such control.

The reasons for this concern vary. The concentration of control of patents in agricultural biotechnology raises concerns for some farmers and consumers that multinational corporations want to simplify the regulatory processes of governments to gain easier approval for their products without adequate review. Because so much of the technology is held by private companies, some regulators and researchers are concerned about how difficult it can be to obtain necessary information for appropriate regulation. Others are concerned that large companies can conduct field experiments, sell seeds to farmers, and market genetically engineered products without appropriate attention to tests for the safety of consumers and the environment. Critics want more testing before farmers expand their planted acreage. There is uncertainty about how much testing is required, whether it should focus on the new genes themselves or on the substances they produce in the plant. Federal agencies can ensure adequate safety testing of biotechnology products by requiring companies to review impacts or by conducting tests for themselves. The FDA proposed rule, discussed earlier, could mandate that industry submit this data. However, others feel that independent testing is necessary to validate the safety of the food product.

There also appears to be concern about how much control companies wield over research in food and agricultural biotechnology. Companies hold patents and own specific germplasm and research techniques needed for plant research. Some companies claim partial ownership of a food product created using their patented technologies. Biotechnology researchers have raised concerns that some private company scientists are not permitted to share innovations in research.

Most food biotechnology research is financed with private, not public, money, and the total amount spent is confidential. One newspaper reported that Monsanto estimates that research and development time and costs to create a commercial product are about 10 years and about $300 million. For every genetically engineered seed that goes to field trials, 10,000 have failed along the way. Corporations charge steep prices for this technology, claiming the need to recoup their investment to be able to research the third and fourth wave of products. Most identifiable public funding for food biotechnology research for plant genome research is being sponsored by the U.S. Department of Agriculture (USDA). The Clinton Administration did not track federal food and agricultural biotechnology funding of research as a line item in federal budget analyses. Consequently, the total amount of public funding for this research is unclear. Public research funding under the authority of the Bayh-Dole Act of 1985 allows universities receiving grants for plant genomics research to hold the intellectual property rights for any useful discoveries. The Act has accelerated the linkages between university research and the creation of consumer products and contributed to the international competitiveness of U.S. industries. It has also encouraged research on minor-use crops.

International Trade Issues

At the present time, the United States is leading the world in biotechnological research, development of genetically modified organisms (GMO's as they are called internationally), and sales of the technology worldwide. The United States has no immediate challengers to this trading position. Some trade experts suggest that trading partners whose policies strongly reflect consumer concerns about the new technology are merely attempting to allow their own domestic industry time to develop a competitive position in this trade. For example, the European Union (EU) has now required that all GMOs be labeled as such, and has severely limited imports of food containing genetically engineered crops and foods. The United States, however, claims that there is not scientific basis to presuppose that genetically modified food products are more risky or substantially different from other products.

U.S. officials believe that decisions on trade should be science-based and U.S. regulatory policy reflects this thinking. Such competition for leadership in biotechnology has influenced trade relationships among U.S trading partners. Japan has diminished its import of U.S. corn after finding StarLink in their imports.

Recombinant DNA technology is producing revolutionary changes in agriculture. Supporters of this technology emphasize the potential benefits from these changes including the promise of higher yields and nutritionally enhanced foods from genetically modified commodities with reduced environmental impact. Opponents of the use of rDNA technology in agriculture are concerned about possible hazards to human, animal, and environmental health. They advocate more safety testing and labeling of products. Many are concerned about the possible consolidation and control of agricultural biotechnology by a handful of multinational corporations. Since the technology potentially can provide greater yields, farmers are rapidly planting genetically modified seeds. As the U.S. experience with this new technology grows, it may become necessary to amend regulatory policies to protect public safety and the environment while allowing genetically modified crop development to progress. The agencies involved in regulating genetically modified foods (FDA, APHIS, and EPA) are implementing policies based upon a 1986 framework document that coordinates their regulatory activities for biotechnology products. This framework applies the same set of regulations to all food products and does not differentiate between foods that are produced with rDNA technologies and those that are produced by traditional methods. However, with the January 18, 2001 proposed rule, making mandatory the current voluntary consultation process whereby the industry now must formally notify the agency 120 days before releasing a food product onto the market. The proposed rule also lists the safety test data that must accompany the notification.

Although the United States is leading the world in the production and sales of genetically engineered products, a few foreign countries have resisted or diminished their U.S. commodity imports because of concerns over the safety of these bioengineered crops. Some suggest that foreign countries' resistance to genetically engineered crops can be traced to their desire to allow their domestic industry time to develop a competitive position in this trade and the growing unease in international trade relationships over the fast adoption by U.S. farmers of bioengineered crops.

Others argue that Congress could exercise more oversight over the regulation of these food crops, fund more public research, and encourage the Administration to negotiate the easing of trade barriers and harmonizing standards. Congress will be closely monitoring these events. The use of biotechnology to produce genetically engineered foods can potentially provide greater yields of nutritionally enhanced foods from less land with reduced use of pesticides and herbicides. This technology has both critics and supporters. Concerns presented to Congress include potential detrimental effects to human and animal health and the environment, and violation of religious customs. Supporters, including individual companies, trade organizations, scientific professional societies, and academic groups, promote benefits such as enhanced crop yields, better nutritional content in food, less pesticide use, and greater agricultural efficiency. They want Congress to defend the U.S. competitive position in export trade of food biotechnology products. Calls for "right-to-know" labeling or other federal regulatory requirements, on the other hand, spark concerns about possibly impeding innovation and adding costs.

In the United States, the regulation of biotechnology food products does not differ fundamentally from regulation of conventional food products. Three federal agencies are primarily responsible for the regulation of genetically engineered foods — the Food and Drug Administration (FDA), the Environmental Protection Agency (EPA), and the U.S. Department of Agriculture (USDA). Each federal agency is assigned certain regulatory responsibilities. FDA provides voluntary pre-market consultations with food companies, seed companies, and plant developers (which they propose to make mandatory) to ensure that biotechnology derived foods meet regulatory standards for safety. USDA's Animal and Plant Health Inspection Service (APHIS) licenses field testing of crops prior to commercial release of newly developed plant strains. EPA registers pesticides in U.S. commerce (including plants engineered to produce pesticides) and establishes levels at which pesticides in foods are permitted. The White House outlined this multi-agency approach to regulating the products of biotechnology in a 1986 document entitled Coordinated Framework for Regulation of Biotechnology. Some critics are concerned about genes from genetically modified plants escaping into the environment through cross fertilization. Others fear the potential overuse of Bt, a natural insecticide, will cause insects to develop resistance to its toxic effects. They want more testing on the long-term environmental and health impact of crops that are altered to produce it. Plus they want the government and academic institutions to conduct independent testing to verify industry data. Industry groups, however, contend that current regulations more than adequately ensure human health and safety. The United States is leading the world in privately funded biotechnological research, genetically modified products, and sales of the technology. During the last days of the Clinton Administration, FDA published a proposed rule changing the way the agency currently regulates bioengineered foods, and published a draft of a guidance document that will advise the biotechnology industry on how to label its food products. In addition, EPA published final rules on how the agency will regulate bioengineered crops, called pest-incorporated protectants.

BIOTECHNOLOGY: FINDING A PRACTICAL APPROACH TO A PROMISING TECHNOLOGY

Biotechnology is emerging as one of the most controversial issues in the global economy. A poorly informed and at times emotional public debate risks disrupting global agricultural trade, inflaming transatlantic relations, and stifling the development of a promising technology. It is important to have thoughtful discussions about this issue and for responsible voices to

be heard. Let's begin by defining terms. Biotechnology in agriculture is a collection of scientific techniques, including genetic engineering, that are used to improve or modify plants and microorganisms. The scientific foundations of biotech agriculture include not only basic plant biology but the knowledge gained in recent decades on the information contained in DNA and the particular functions of genetic material in nature. Traditional crossbreeding has been practiced around the world for hundreds of years as a way to improve agriculture. When I was a youngster growing up in Iowa, any drive through the country inevitably would pass by fields of hybrid corn. There was never anything mysterious about it. Readers from rural areas anywhere in the world likely have had similar experiences. Modern biotechnology has dramatically improved on traditional crossbreeding by permitting a much more targeted approach to genetic modification. It offers solutions to longstanding problems, such as pest damage, that rely on more precise approaches. Biotech allows minute genetic changes that can make a plant unappealing to a specific pest, for example, and may render unnecessary broad spectrum pesticides that negatively affect both water and soil, and even damage the health of people who work on farms. Corn and soybean varieties produced with the help of modern biotechnology already are permitting reduced use of herbicides and pesticides, with important benefits to the environment.

Biotechnology has the potential to make agriculture much more productive. This is an immensely important consideration in a world where millions die of starvation, where population growth continues to place great stress on food production capacity, where there is a pressing need to avoid cultivation of environmentally fragile land, and where many families still rely on farming for their livelihood. Biotechnology agriculture is also therefore a useful tool in our efforts to fight hunger and poverty.

Trade in Biotechnology Products

Scientists believe that biotech has the potential to increase crop yields by 20 percent or more with no greater use of natural resources, even on small farms. Developing country farmers and consumers should have access to the same agricultural choices and benefits that the rest of the world does. This promising beginning has been marred already by Europe's refusal to permit imports of biotech corn varieties approved by competent European authorities.

As a result, American corn producers are now losing some $200 million in legitimate exports annually. An even larger concern is the risk of serious and broader disruption of about a half a trillion dollars in annual agricultural trade worldwide. These trade flows help ensure adequate food supplies at reasonable prices for people everywhere.

The current situation has already caused a worrisome strain in transatlantic trade relations and has the potential to affect trade relations worldwide. Additional measures by Europe would open the possibility of opening a far more serious transatlantic rift, at precisely the time when Europe and the United States need to be working together and with others to launch a new trade round. And the new round is essential to spurring global growth, which translates into new jobs at higher wages and resources to improve living standards for people everywhere.

Assuring Food Safety

American farmers understand that some consumers in Europe and elsewhere have sincerely felt questions and concerns about biotech food. As governments and citizens address these issues, it is important to be guided by what science actually tells us about the safety of these products. I have had to look carefully myself at what the scientists have found. The overwhelming majority of scientific experts worldwide — both private and public — based on years of research, believe that biotech foods are safe for people to eat. In fact, all the evidence that we have — and it is considerable — indicates that biotech foods are as safe as conventional foods, even those foods that have been around for hundreds of years. If some consumers find them unappealing that is another matter, but not one that justifies government-mandated trade barriers or food safety-related restrictions.

Readers should be clear on one point. Food safety is a paramount, overriding concern for Americans. Americans eat food produced through the aid of biotechnology. In fact, they eat lots of it. If there were scientific evidence that biotech food posed a threat to human health, such food would not be on the market in the United States. Consumer acceptance of biotech food in the United States is largely grounded in our experience with it and in the credibility of the U.S. food safety program. Biotech companies do extensive testing of all new crops before marketing them. They consult with the Food and Drug Administration (FDA) on any safety or regulatory questions. The FDA has been consulted on every biotech food that is now on the market.

Some of my European friends, while not debating the science, defend blocking imports of biotech food on cultural grounds, perhaps because they believe they have higher standards for purity and safety. Americans, they argue, somehow take less interest in the safety of the food they eat than do Europeans. As a diplomat who recently spent three wonderful years in France, I find this argument hard to swallow. To be sure, some Europeans still buy food daily, prepare it lovingly, and make meals a leisurely, social occasion in a manner. But these practices, however admirable, do not speak to the question of concern over food safety.

Americans, in fact, despite our penchant for so-called "fast foods," are rather fussy about the safety of the things we ingest. We have insisted on very strict rules on the use of food additives. We have created scientifically grounded, apolitical, and respected entities like the FDA to oversee food safety. American shoppers inspect labels of food products to avoid foods that contain cholesterol, sugar, and saturated fat, substances that have been scientifically shown to pose health risks for some people.

And, of course, tobacco products contain strong warning labels in light of their demonstrated link to lung cancer and heart disease. In Europe, I did not see greater, or even the same, level of concern about substances people ingest into their bodies. It is telling that Europeans still smoke much more than do Americans, despite what we know from science about the damaging effects to human health. There is also still no effective, Europe-wide food safety regulatory body. Some national European food safety agencies, however, have an unfortunate record of being susceptible to political influence. Politicians in Europe sometimes have slowed the removal of contaminated food from supermarkets. And as many have pointed out, governmental credibility on food safety was eroded when some European political leaders dismissed the seriousness of "mad cow" disease. All things considered, I see no evidence that ordinary citizens from Europe or anywhere else are more concerned than the ordinary American about what they put into their bodies. The bottom line is that all of us have an interest in assuring a safe food supply in a manner that allows this new technology to achieve its potential.

A More Productive Agricultural Sector

It is important to understand that biotechnology has enormous potential benefits. Not least among these benefits is the potential to reduce the environmental impact of agriculture. Some biotech crops can decrease the need for pesticides and herbicides to control pests, weeds, and plant diseases and allow more selective application of agricultural
8 chemicals. For example, biotech potato, corn, and cotton plants have been engineered to produce a bacterial toxin (*Bacillus thuringiensis*, or Bt delta endotoxin) found in natural soils to ward off destructive insects on their own. And biotech herbicide-tolerant cotton, corn, and soybeans have given farmers a choice to make fewer applications of

herbicides or to use more environmentally friendly herbicides.

Biotech has as well the potential to provide enhanced resistance to variations in temperature, soil salinity, and the availability of water. For example, plants can potentially be enhanced to withstand a drop in temperature and frost by modifying their production of linoleic acid. Even the land's ability to support continued farming can potentially be supported through the use of herbicidetolerant biotech crops that require less tillage to reduce plant pests. Cutting back on tillage lessens both soil and water runoff and soil nutrient depletion. Scientists are also looking at ways to use biotech to deliver more nutrients and better taste in our foods. Damaging deficiencies in Vitamin A and other nutrients among the poor worldwide may well be addressed costeffectively through biotech agriculture. In a nonagricultural application, scientists are also looking at how biotech might render mosquitoes unable to transmit malaria. Millions of lives could be saved and the use of highly toxic chemicals against mosquitoes reduced or done away with. Another potential benefit of biotech is increased income for farmers, both small and large. For example, biotechnology has improved the quality of seed grains and the ability to produce bigger harvests from currently cultivated land. Equally important, increased yields and reduced chemical and labor costs can represent increased income for the farmer. Finally, farmers can save in the cost of bringing their product to market with biotech crops that require less handling, are easier to store, need no refrigeration, and have a longer shelf-life.

Promoting a Global Dialogue

Views on biotech around the world are still being formed and influenced. There are important questions to address and differing viewpoints to examine. What is the best way forward in pursuit of a consensus? I would suggest the following points:

1. Cool the Rhetoric. It can be difficult to have a rational public debate on a question that is both scientific and emotionally charged. It is troubling to see slogans such as "Frankenstein food" substituting for intelligent discussion. It makes no sense to consider new scientific developments as "Frankenstein" and existing science as "natural." In a free society, those who engage in irresponsible sloganeering will find their speech securely protected but rarely respected.

2. Don't Take Precipitous Actions. People and governments everywhere are struggling to ensure that rapid technological changes in an increasingly integrated world bring widely shared benefits. With biotech, as with other important breakthroughs, this needs to be done on the basis of sound information and science. It requires an objective effort to safeguard against real risks while making benefits and choices available to everyday people. On complex issues, a rush to judgment seldom allows for such a process. As Romano Prodi, the new president of the European Commission, said recently, "we want to avoid knee-jerk reactions" on food safety issues and "not rush in foolishly."

3. Restore Predictability. The world's farmers are the backbone of our food supply. They and all those people and companies involved with agriculture and food deserve to know the rules under which they must operate, and to have the criteria and application of those rules be predictable and transparent. Agricultural planning decisions and investments are made long in advance of when products hit the market. Governments have the primary responsibility to bring predictability to food safety regulations and to create consumer confidence through apolitical, technical implementation processes that are themselves predictable.

4. Use Established Multilateral Discussion Fora. We already have the institutions and mechanisms for international discussions of biotechnology. Chief among them is the Codex Alimentarius, an organization jointly sponsored by the U.N. Food and Agriculture Organization (FAO) and the World Health Organization (WHO). The Codex, comprised of independent scientists from around the world, is not a very flashy or politically powerful organization. Nonetheless, it has for many years kept the international community well informed of the scientific basis for food safety considerations and the implications in the international arena. Codex's strength lies in its broad membership, its extensive, international technical capacity, and the fact that it deals at arms-length from the policy process of regions and countries. It would be impossible to duplicate what we have already created in the Codex, in protection for our consumers and support for an open, international economy based on fair rules of the game.

5. Stress Science-based Approaches. We must be acutely aware of protecting the scientific basis for decisions that can block trade, stifle innovation, feed isolationism and fear, and deny our citizens the right to choose new products out of nostalgia or our own discomfort. Sound science, continually applied, remains the best foundation for food-safety decisions and practices.

6. Address Public Questions. The public's curiosity and fears are being aroused by misleading and partial stories about biotech foods and about the science behind them. It is important to respond to those concerns and to facilitate an open exchange of information, particularly well-grounded information. In a cynical age, it is hard to combat lightening-quick "one-liners" and Internet tales, but there is still a difference between information and spin. Science is by no means perfect, but the information on the safety of biotech

foods we have gotten from careful, scientific analysis, properly reviewed and established over time, should be shared with everyday consumers. Even if it is unpopular, we should be straight with consumers that extensive analysis has turned up no special health risks from biotech foods. As one scientist put it, there are substantial known risks to a high-fat diet and a lack of exercise, but you do not see that making the headlines on a regular basis like biotech does.

7. Preserve WTO Rules. The reason all 134 members of the World Trade Organization (WTO) have agreed to abide by its rules is extremely simple. We all know that we are better off with clear rules rather than chaos, and that the rules help prevent every one of us from seeking unfair advantage in the global trading system. The rulesbased trading system may not be perfect and may even occasionally transgress our sense of independence or uniqueness, but they also bind us as a community and enable the incredible choices and benefits of trade we have come to enjoy. There is nothing about biotech food that would make it necessary or worthwhile to abrogate WTO rules. I cannot stress enough how important it is to protect this institution and to strengthen it without damaging its fundamental purpose, which is to facilitate fair trade among all countries.

There are also contact numbers for people within the U.S. government who can help answer your questions. I can assure you that the United States Government will continue to safeguard the safety of its consumers, even as we work with others to shed real light on this issue.

Regulating the Products of Biotechnology

If genetically engineered organisms are to gain greater acceptance, decisions that address concerns associated with the application of biotechnology to agriculture must be science based. Science also must be the base by which regulatory officials can assure and build upon credibility, remain current, and assure a rational basis for decisionmaking. In this way, science and the legal processes are inextricably linked for regulations that evaluate biological products. This article outlines how U.S. regulatory authorities examine biotechnology products and coordinate their evaluation procedures to assure the highest degree of safety.

A Science-Based Regulatory Approach

To say that an assessment is science-based means that the review of any particular product is done using scientific criteria relevant to that product. Advances in biotechnology — being able to determine what has actually happened at the molecular and biochemical levels — have increased the ability of regulators to scrutinize product safety and the effect of product modification upon safety. The scrutiny of certain products and the rigor by which evaluations are performed have also increased due to biotechnology. The approach to review is constantly evolving due to new types of products and the availability of new scientific information.

The current regulatory approach in the United States is based on a determination in 1987 by the U.S. National Academy of Sciences (NAS) that "the risks associated with the introduction of rDNA-engineered organisms are the same in kind as those associated with the introduction into the environment of unmodified organisms and organisms modified by other genetic techniques." In 1989, NAS articulated the Concept of Familiarity, saying "familiar does not necessarily mean safe. Rather, to be familiar with the elements of an introduction [into the environment] means to have enough information to be able to judge the introduction's safety or risk." The familiarity criterion is central to a framework for evaluation of genetically modified plants and microorganisms. For plants, familiarity comes from comparison with the parent line or crop species with similar traits, as well as from actual field tests. With genetically modified plants, the standard of comparison is the unmodified plant.

Approaches to review were developed in the United States by the NAS and, among industrialized countries, by the Organization for Economic Cooperation and Development (OECD). In 1993, the OECD published principles for reviewing large-scale plantings of transgenic crop plants. That organization identified and discussed the general safety issues associated with new plant lines and crop cultivars, as well as risk management. These safety issues, including gene transfer, weediness, and effects on non-target organisms, are now generally accepted globally as the basis for evaluating transgenic plants. The OECD elaborated on the Concept of Familiarity, identifying familiarity with the crop plant, the trait, the environment, and the interaction of these three factors as being critical in the evaluation of transgenic plants.

The U.S. Decision-making Process

Since the White House Office of Science and Technology Policy published the "Coordinated Framework for the Regulation of Products of Biotechnology" in 1986, the United States has gained 13 years' experience in evaluating the products of biotechnology for safety. The Coordinated Framework established a strong federal commitment to the safe development of the products of biotechnology from the laboratory, through field-testing and development, and into the marketplace. The underlying assumption of the Coordinated Framework is that any risks from the products of biotechnology are the same in kind as those of similar products — risks to agriculture, the environment, and human health. Thus, the existing statutes for addressing these risks have been deemed adequate to address any risks posed by product development using biotechnology, and no new "gene law" has been considered necessary.

Regulations, developed to implement statutes, establish procedures and criteria by which specific types of products are evaluated, including those produced using biotechnology such as vaccines, plant varieties, pesticides, animal products, and pharmaceuticals. In the United States, the agencies that examine plants and plant products are the Environmental Protection Agency (EPA), the Food and Drug Administration (FDA) of the Department of Health and Human Services, and the Animal and Plant Health Inspection Service (APHIS) of the U.S. Department of Agriculture.

Under the U.S. regulatory system, a new genetically engineered plant could be reviewed by one or all three regulatory agencies, depending on the plant and the trait engineered into it. For instance, a Bt gene in a food crop would be reviewed by APHIS, EPA, and FDA; a plant with modified oil content for food would be reviewed by FDA and APHIS; modified flower color in a horticultural crop would be reviewed by APHIS alone. It usually takes at least five years of field testing, under APHIS oversight, for the developer of a new plant variety to evaluate the new line and to collect the data needed to pass through the regulatory system. Another two years may be needed for APHIS, EPA, and/or FDA (depending on the plant variety) to complete their reviews. The U.S. regulatory structure is based upon risk rather than process to assure safety, and its success is due to the fact that regulatory agencies with established credibility and expertise have been designated to evaluate the products of biotechnology.

The Regulators' Roles

APHIS regulates the development and field testing of

genetically engineered plants, microorganisms, and certain other organisms under the authority of the Federal Plant Pest Act and the Plant Quarantine Act. APHIS regulations provide procedures for obtaining a permit or for providing notification of the intent to field test, prior to importation, interstate movement, or release into the United States. Authority is based on the use of plant pests to engineer plants so as to supply the genes inserted, to allow those genes to be expressed in plants, or to transfer the genes into plants. The APHIS review process evaluates agricultural and environmental safety issues.

APHIS has been reviewing applications for permits and notifications by industry, academia, and nonprofit organizations of field tests of transgenic crop plants since 1986, when it proposed the first regulation for these products. After several years of field tests, an applicant can petition the agency to be released from requirements under the APHIS regulatory process. If the applicant can provide evidence that there is no plant pest potential (including the lack of change in disease and pest resistance, as well as the absence of the potential of new genetic material to create a new pathogen or pest), along with answers to a variety of other environmental questions, APHIS will grant the petition. At that time, the applicant is free to commercialize the plant line or use it in other breeding programs without coming to APHIS for permission, subject to obtaining any necessary approvals from EPA or FDA. To date, 51 petitions have been granted and more than 5,000 permits and notifications have been issued for field testing at more than 22,000 sites. Although no petitions have been denied, 13 have been withdrawn due to insufficient information or other inadequacies in the application.

APHIS maintains comprehensive field testing and petition databases. These databases are used not only by domestic customers and stakeholders, but increasingly by foreign governments to verify that the U.S. government has reviewed the risk for products being considered for field testing or importation. These databases, as well as access to federal home pages on biotechnology regulation, are available at www.aphis.usda.gov. EPA's responsibility is to ensure the safety of pesticides, both chemical and biological, under the authority of the Federal Insecticide, Fungicide, and Rodenticide Act (FIFRA) through regulation of the distribution, sale, use, and testing of plants and microbes producing pesticidal substances. Under the Federal Food, Drug and Cosmetic Act (FFDCA), EPA sets tolerance limits for substances used as pesticides on and in food and feed, or establishes an exemption from the requirement of a tolerance if such a tolerance is not necessary to protect the public health (determined after evaluation by the agency).

EPA issues experimental use permits for field testing of "pesticidal" plants and registrations for commercialization of these plants. The *Bacillus thuringiensis* (Bt) toxin, which occurs naturally in soil bacterium, is considered a biological pesticide. For plants containing Bt toxin, the manufacturer must prepare a resistance management plan as a condition for registration with EPA. The plan describes how the manufacturer registering the plant product will assure that resistance does not build up in affected insect populations and reduce the effectiveness of Bt applied topically or used through the plant's genetics.

EPA also evaluates the new use of herbicides on herbicide-tolerant transgenic plants. FDA assesses food (including animal feed) safety and nutritional aspects from new plant varieties as part of a consultation procedure published in the 1992 Statement of Policy: Foods Derived From New Plant Varieties. Consistent with its 1992 policy, FDA expects developers of new plant varieties to consult with the agency on safety and regulatory questions under the authority of the FFDCA. FDA policy is based on existing food law and requires that genetically engineered foods meet the same rigorous safety standards as is required of all other foods.

The FDA biotechnology policy treats substances intentionally added to food through genetic engineering as food additives if they are significantly different in structure, function, or amount from substances currently found in food. Many of the food crops currently being developed using biotechnology do not contain substances that are significantly different from those already in the diet and thus do not require pre-market approval.

Science informs the decision-making process of the regulators at a variety of levels. The available published scientific literature, particularly from peer-reviewed journals, is used by regulators in evaluating specific products. Applicants cite this literature in their applications for regulatory approval. As in the case of the NAS, the scientific enterprise can be asked to identify the scientific issues and recommend approaches to particular types of products. Meetings of scientists can be called to address specific issues, as have past meetings on Bt, viral recombination, and relevant biological factors for evaluating crop plants. Information can even be requested on specific products. The EPA meets with its scientific advisory panels. The FDA refers questions to its Food Advisory Committee. Recently, U.S. Agriculture Secretary Dan Glickman requested that the NAS review the scientific underpinnings of the regulatory decisions made by USDA. The USDA also has a Risk Assessment Grants Program that specifically funds research on emerging issues with genetically engineered organisms. All of this information is used by regulators to assure that the most current information and thinking is available to inform regulatory decisions.

Beyond the U.S.

Internationally, the appropriate scientifically based standards, guidelines, and recommendations for transgenic products and other products as they move into the international marketplace, are being developed by the representatives of national governments at working groups and task forces devoted to these issues under the standard-setting bodies — Codex Alimentarius, the International Plant Protection Convention, and Office des International Epizooties. The use of the existing standard-setting bodies to address concerns about products of biotechnology focuses attention on harmonization of risk-assessment methodologies and evaluation of specific products or classes of products.

3

Biotechnology Policy, Guidelines and Statutes

The advent of a technology like genetic engineering poses important challenges to society. How should decisions be made about which applications of the technology should go forward and which should not? Current US institutions provide meager opportunities to openly evaluate new technologies in advance of their implementation. As a result, decisions about technologies are largely left to the private sector, which is given full latitude to produce whatever can be sold, subject only to regulation to avoid harm to health and the environment. While such regulation is important, it is no substitute for societal evaluation of the impacts of and alternatives to new technologies. This timid approach to technology has led to an impression that it is somehow beyond social or political control. Yet choices between different technologies or between technological and nontechnological approaches to problems are being made all the time. Often the decisions are made in the context of funding decisions and the setting of research agendas, which occur far from public view. New institutions and arenas are needed to encourage open discussion of new technologies and to help the public participate in decisions about technologies.

US Regulation of Genetic Engineering Products

Three agencies currently regulate genetically engineered products under a handful of statutes. These statutes were originally designed to regulate other products and were put in place before genetically engineered products had even been imagined. Thus they have been adapted to a purpose for which they were not intended. The result is a patchwork of regulations and programs with different requirements and underlying philosophies.

In general, the products of genetic engineering are regulated for their health and environmental risks, and those risks are balanced against benefits. The major statutes applied to nonpharmaceutical and nontherapeutic uses of genetic engineering are implemented by three agencies: the Environmental Protection Agency, the Food and Drug Administration, and the United States Department of Agriculture. The Environmental Protection Agency (EPA) oversees genetically engineered microbial pesticides and certain genetically engineered crops under the Federal Insecticide, Fungicide, and Rodenticide Act (FIFRA) and nonpesticidal, nonfood microbial products under the Toxic Substances Control Act (TSCA).

Organisms regulated under FIFRA are subject to stringent permit standards under which the EPA can require new studies on possible harm to health and the environment. By contrast, the legal provisions of TSCA require only 90-days notice to the EPA before the commencement of manufacture and provide less straightforward mechanisms for persuading companies to conduct studies of risks. Currently, the EPA regulates Bt crops and microbial pesticides under FIFRA and microorganisms intended for pollution cleanup under TSCA. The Food and Drug Administration (FDA) has the authority to regulate all genetically engineered foods under the Food, Drug and

Cosmetic Act. Its only specific policy for genetically engineered foods applies to crops. Under the current FDA food policy, most genetically engineered crops are not subject to regulation. The agency does run a voluntary program under which companies producing engineered crops can consult with the agency to confirm the regulatory status of their products. Companies choosing to participate in the consultation process are not required to conduct any standard set of safety tests on their products. In fact, companies that have voluntarily conducted safety studies only need to submit summaries of those studies to the FDA. Because the FDA reviews only summaries, not data, from safety studies, it does not publish conclusions on the safety of individual genetically engineered foods. The FDA has conducted consultations on genetically engineered tomatoes, soybeans, and squash.

The US Department of Agriculture (USDA) regulates genetically engineered products under several statutes: genetically engineered crops under the Plant Pest Act and related authorities; animal vaccines under the Virus, Serum and Toxin Act; and engineered poultry and livestock under various meat inspection statutes. Under the Plant Pest Act, the USDA issues de facto commercial permits in the somewhat confusing form of determinations that the products are not pests and do not need further regulation. These determinations are accompanied by environmental assessments done under the National Environmental Policy Act (NEPA). The USDA Plant Pest Act program has overseen thousands of field tests of genetically engineered crops and issued about twenty de facto commercial permits. Crops approved for commercial use include engineered versions of most major commodity crops such as corn, cotton, and soybeans, as well as several fruits and vegetables. In general, genetically engineered animals other than poultry and livestock are not regulated for environmental risks under US law. Engineered fish are a notable example. There is some indication that the FDA may regulate engineered fish as "animal drugs," but no official policy has been announced.

The Need for a New Regulatory Framework

The framework of regulations applied to genetic engineered organisms is weak because it leaves important categories of genetically engineered organisms uncovered and because it often lacks strong authority to require premarket review of genetically engineered products. The current programs differ widely in the degree to which they allow the public to participate in decisions about engineered products. Most of the statutes grant companies broad privileges to withhold data and information, including health and safety information, from the public. Making these statutes work requires active involvement by the public. The agencies mentioned above almost always hear from the regulated industry on pending policies and decisions. Unless agencies hear from the public too, their positions often coincide with those of industry. Decisions about genetic engineering should not be left to the multinational corporations developing these products. Letting the agencies know that the public cares about genetic engineering is the best way to positively influence their decisions.

BUSH ADMINISTRATION NIH GUIDELINES FOR EMBRYONIC STEM CELL FUNDING TAKEN FROM THE WEBSITE OF THE NATIONAL INSTITUTES OF HEALTH ON SEPTEMBER 26, 200 .

The text follows.

Notice of criteria for federal funding of research on existing human embryonic stem cells and establishment of nih human embryonic stem cell registry

Release Date: November 7, 2001

NOTICE: NOT-OD-02-005

Office of the Director, NIH

On August 9, 2001, at 9:00 p.m. EDT, the President announced his decision to allow Federal funds to be used for research on existing human embryonic stem cell lines as long as prior to his announcement (1) the derivation process (which commences with the removal of the inner cell mass from the blastocyst) had already been initiated and (2) the embryo from which the stem cell line was derived no longer had the possibility of development as a human being.

In addition, the President established the following criteria that must be met:

(i) The stem cells must have been derived from an embryo that was created for reproductive purposes;
(ii) The embryo was no longer needed for these purposes;
(iii) Informed consent must have been obtained for the donation of the embryo;
(iv) No financial inducements were provided for donation of the embryo.

In order to facilitate research using human embryonic stem cells, the NIH is creating a Human Embryonic Stem Cell Registry that will list the human embryonic stem cells that meet the eligibility criteria. Specifically, the laboratories or companies that provide the cells listed on the Registry will have submitted to the NIH a signed assurance. Each provider must retain for submission to the NIH, if necessary, written documentation to verify the statements in the signed assurance.

The Registry will be accessible to investigators on the NIH Home Page http://escr.nih.gov. Requests for Federal funding must cite a human embryonic stem cell line that is listed on the NIH Registry.

Such requests will also need to meet existing scientific and technical merit criteria and be recommended for funding by the relevant National Advisory Council, as appropriate. Further guidance is accessible at http://grants.nih.gov/ grants/ guide/ notice-files/NOTOD- 02-006.html.

CLINTON ADMINISTRATION NIH GUIDELINES FOR EMBRYONIC STEM CELL FUNDING

As printed in the Federal Register, August 25, 2000 ("National Institutes of Health Guidelines for Research Using Human Pluripotent Stem Cells," 65 Fed. Reg. 51,975, Aug. 25, 2000) The text of the final Guidelines follows.

I. Scope of Guidelines

These Guidelines apply to the expenditure of National Institutes of Health (NIH) funds for research using human pluripotent stem cells derived from human embryos (technically known as human embryonic stem cells) or human fetal tissue (technically known as human embryonic germ cells). For purposes of these Guidelines, "human pluripotent stem cells" are cells that are self-replicating, are derived from human embryos or human fetal tissue, and are known to develop into cells and tissues of the three primary germ layers. Although human pluripotent stem cells may be derived from embryos or fetal tissue, such stem cells are not themselves embryos. NIH research funded under these Guidelines will involve human pluripotent stem cells derived: (1) From human fetal tissue; or (2) from human embryos that are the result of in vitro fertilization, are in excess of clinical need, and have not reached the stage at which the mesoderm is formed.

In accordance with 42 Code of Federal Regulations (CFR) 52.4, these Guidelines prescribe the documentation and assurances that must accompany requests for NIH funding for research using human pluripotent stem cells from: (1) Awardees who want to use existing funds; (2) awardees requesting an administrative or competing supplement; and (3) applicants or intramural researchers submitting applications or proposals. NIH funds may be used to derive human pluripotent stem cells from fetal tissue. NIH funds may not be used to derive human pluripotent stem cells from human embryos. These Guidelines also designate certain areas of human pluripotent stem cell research as ineligible for NIH funding.

II. Guidelines for Research Using Human Pluripotent Stem Cells That Is Eligible for NIH Funding A. Utilization of Human Pluripotent Stem Cells Derived From Human Embryos

1. Submission to NIH: Intramural or extramural investigators who are intending to use existing funds, are requesting an administrative supplement, or are applying for new NIH funding for research using human pluripotent stem cells derived from human embryos must submit to NIH the following:

a. An assurance signed by the responsible institutional official that the pluripotent stem cells were derived from human embryos in accordance with the conditions set forth in section II.A.2 of these Guidelines and that the institution will maintain documentation in support of the assurance;

b. A sample informed consent document (with patient identifier information removed) and a description of the informed consent process that meet the criteria for informed consent set forth in section II.A.2.e of these Guidelines;

c. An abstract of the scientific protocol used to derive human pluripotent stem cells from an embryo;

d. Documentation of Institutional Review Board (IRB) approval of the derivation protocol;

e. An assurance that the stem cells to be used in the research were or will be obtained through a donation or through a payment that does not exceed the reasonable costs associated with the transportation, processing, preservation, quality control and storage of the stem cells;

f. The title of the research proposal or specific subproject that proposes the use of human pluripotent stem cells;

g. An assurance that the proposed research using human pluripotent stem cells is not a class of research that is ineligible for NIH funding as set forth in section III of these Guidelines; and

h. The Principal Investigator's written consent to the disclosure of all material submitted under Paragraph A.1 of this section, as necessary to carry out the public review and other oversight procedures set forth in section IV of these Guidelines.

2. Conditions for the Utilization of Human Pluripotent Stem Cells Derived From Human Embryos: Studies utilizing pluripotent stem cells derived from human embryos may be conducted using NIH funds only if the cells were derived (without Federal funds) from human embryos that were created for the purposes of fertility treatment and were in excess of the clinical need of the individuals seeking such treatment.

a. To ensure that the donation of human embryos in excess of the clinical need is voluntary, no inducements, monetary or otherwise, should have been offered for the donation of human embryos for research purposes. Fertility clinics and/or their affiliated laboratories should have implemented specific written policies and practices to ensure that no such inducements are made available.

b. There should have been a clear separation between the decision to create embryos for fertility treatment and the decision to donate human embryos in excess of clinical need for research purposes to derive pluripotent stem cells. Decisions related to the creation of embryos for fertility treatment should have been made free from the influence of researchers or investigators proposing to derive or utilize human pluripotent stem cells in research. To this end, the attending physician responsible for the fertility treatment and the researcher or investigator deriving and/or proposing to utilize human pluripotent stem cells should not have been one and the same person.

c. To ensure that human embryos donated for research were in excess of the clinical need of the individuals seeking fertility treatment and to allow potential donors time between the creation of the embryos for fertility treatment and the decision to donate for research purposes, only frozen human embryos should have been used to derive human pluripotent stem cells. In addition, individuals undergoing fertility treatment should have been approached about consent for donation of human embryos to derive pluripotent stem cells only at the time of deciding the disposition of embryos in excess of the clinical need.

d. Donation of human embryos should have been made without any restriction or direction regarding the individual(s) who may be the recipients of transplantation of the cells derived from the human pluripotent stem cells.

e. Informed Consent Informed consent should have been obtained from individuals who have sought fertility treatment and who elect to donate human embryos in excess of clinical need for human pluripotent stem cell research purposes. The informed consent process should have included discussion of the following information with potential donors, pertinent to making the decision whether or not to donate their embryos for research purposes. Informed consent should have included:

(i) A statement that the embryos will be used to derive human pluripotent stem cells for research that may include human transplantation research;

(ii) A statement that the donation is made without any restriction or direction regarding the individual(s) who may be the recipient(s) of transplantation of the cells derived from the embryo;

(iii) A statement as to whether or not information that could identify the donors of the embryos, directly or through identifiers linked to the donors, will be removed prior to the derivation or the use of human pluripotent stem cells;

(iv) A statement that derived cells and/or cell lines may be kept for many years;

(v) Disclosure of the possibility that the results of research on the human pluripotent stem cells may have commercial potential, and a statement that the donor will not receive financial or any other benefits from any such future commercial development;

(vi) A statement that the research is not intended to provide direct medical benefit to the donor; and

(vii) A statement that embryos donated will not be transferred to a woman's uterus and will not survive the human pluripotent stem cell derivation process.

f. Derivation protocols should have been approved by an IRB established in accord with 45 CFR 46.107 and 46.108 or FDA regulations at 21 CFR 56.107 and 56.108.

B. Utilization of Human Pluripotent Stem Cells Derived From Human Fetal Tissue

1. Submission to NIH: Intramural or extramural investigators who are intending to use existing funds, are requesting an administrative supplement, or are applying for new NIH funding for research using human pluripotent stem cells derived from fetal tissue must submit to NIH the following:

a. An assurance signed by the responsible institutional official that the pluripotent stem cells were derived from human fetal tissue in accordance with the conditions set forth in section II.A.2 of these Guidelines and that the institution will maintain documentation in support of the assurance;

b. A sample informed consent document (with patient identifier information removed) and a description of the informed consent process that meet the criteria for informed consent set forth in section II.B.2.b of these Guidelines;

c. An abstract of the scientific protocol used to derive human pluripotent stem cells from fetal tissue;
d. Documentation of IRB approval of the derivation protocol;
e. An assurance that the stem cells to be used in the research were or will be obtained through a donation or through a payment that does not exceed the reasonable costs associated with the transportation, processing, preservation, quality control and storage of the stem cells;
f. The title of the research proposal or specific subproject that proposes the use of human pluripotent stem cells;
g. An assurance that the proposed research using human pluripotent stem cells is not a class of research that is ineligible for NIH funding as set forth in section III of these Guidelines; and
h. The Principal Investigator's written consent to the disclosure of all material submitted under Paragraph B.1 of this section, as necessary to carry out the public review and other oversight procedures set forth in section IV of these Guidelines.

2. Conditions for the Utilization of Human Pluripotent Stem Cells Derived From Fetal Tissue

a. Unlike pluripotent stem cells derived from human embryos, DHHS funds may be used to support research to derive pluripotent stem cells from fetal tissue, as well as for research utilizing such cells. Such research is governed by Federal statutory restrictions regarding fetal tissue research at 42 U.S.C. 289g-2(a) and the Federal regulations at 45 CFR 46.210. In addition, because cells derived from fetal tissue at the early stages of investigation may, at a later date, be used in human fetal tissue transplantation research, it is the policy of NIH to require that all NIH-funded research involving the derivation or utilization of pluripotent stem cells from human fetal tissue also comply with the fetal tissue transplantation research statute at 42 U.S.C. 289g-1.

b. Informed Consent: As a policy matter, NIH-funded research deriving or utilizing human pluripotent stem cells from fetal tissue should comply with the informed consent law applicable to fetal tissue transplantation research (42 U.S.C. 289g-1) and the following conditions. The informed consent process should have included discussion of the following information with potential donors, pertinent to making the decision whether to donate fetal tissue for research purposes. Informed consent should have included:

(i) A statement that fetal tissue will be used to derive human pluripotent stem cells for research that may include human transplantation research;

(ii) A statement that the donation is made without any restriction or direction regarding the individual(s) who may be the recipient(s) of transplantation of the cells derived from the fetal tissue;

(iii) A statement as to whether or not information that could identify the donors of the fetal tissue, directly or through identifiers linked to the donors, will be removed prior to the derivation or the use of human pluripotent stem cells;

(iv) A statement that derived cells and/or cell lines may be kept for many years;

(v) Disclosure of the possibility that the results of research on the human pluripotent stem cells may have commercial potential, and a statement that the donor will not receive financial or any other benefits from any such future commercial development; and

(vi) A statement that the research is not intended to provide direct medical benefit to the donor.

c. Derivation protocols should have been approved by an IRB established in accord with 45 CFR 46.107 and 46.108 or FDA regulations at 21 CFR 56.107 and 56.108.

III. Areas of Research Involving Human Pluripotent Stem Cells That Are Ineligible for NIH Funding

Areas of research ineligible for NIH funding include:

A. The derivation of pluripotent stem cells from human embryos;

B. Research in which human pluripotent stem cells are utilized to create or contribute to a human embryo;

C. Research utilizing pluripotent stem cells that were derived from human embryos created for research purposes, rather than for fertility treatment;

D. Research in which human pluripotent stem cells are derived using somatic cell nuclear transfer, i.e., the transfer of a human somatic cell nucleus into a human or animal egg;

E. Research utilizing human pluripotent stem cells that were derived using somatic cell nuclear transfer, i.e., the transfer of a human somatic cell nucleus into a human or animal egg;

F. Research in which human pluripotent stem cells are combined with an animal embryo; and

G. Research in which human pluripotent stem cells are used in combination with somatic cell nuclear transfer for the purposes of reproductive cloning of a human.

IV. Oversight

A. The NIH Human Pluripotent Stem Cell Review Group (HPSCRG) will review documentation of compliance with the Guidelines for funding requests that propose the use of human pluripotent stem cells. This working group will hold public meetings when a funding request proposes the use of a line of human pluripotent stem cells that has not been previously reviewed and approved by the HPSCRG.

B. In the case of new or competing continuation (renewal) or competing supplement applications, all applications shall be reviewed by HPSCRG and for scientific merit by a Scientific Review Group. In the case of requests to use existing funds or applications for an administrative supplement or in the case of intramural proposals, Institute or Center staff should forward material to the HPSCRG for review and determination of compliance with the Guidelines prior to allowing the research to proceed.

C. The NIH will compile a yearly report that will include the number of applications and proposals reviewed and the titles of all awarded applications, supplements or administrative approvals for the use of existing funds, and intramural projects.

D. Members of the HPSCRG will also serve as a resource for recommendations to the NIH with regard to any revisions to the NIH Guidelines for Research Using Human Pluripotent Stem Cells and any need for human pluripotent stem cell policy conferences.

LEGISLATORS AS LOBBYISTS: PROPOSED STATE REGULATION OF EMBRYONIC STEM CELL RESEARCH, THERAPEUTIC CLONING AND REPRODUCTIVE CLONING

The issue of research involving embryos and fetuses has been a matter of serious social concern for the past three decades. As each new scientific and medical development is introduced that affects embryos and fetuses, new state laws are proposed to deal with them. In the wake of the 1973 U.S. Supreme Court abortion decision in Roe v. Wade, for example, 24 states passed laws prohibiting research on human conceptuses. When *in vitro* fertilization (IVF) was accomplished, at least one state passed a law attempting to discourage the procedure. After Dolly the cloned sheep was born, nine states adopted laws restricting human reproductive cloning. When researchers began creating and using human embryos for stem cell research, states began introducing laws to deal with that technology as well. In 2002 and 2003, dozens of bills were introduced in state legislatures on cloning and stem cell research. In total, 38 states considered such bills; some states considered multiple and conflicting bills. In all, 133 separate bills were introduced on these subjects.

Although state legislatures have addressed regulation of research on the unborn for 30 years, the more recent technologies are being debated within a new context. Pro-life and pro-choice advocates are aligning in unusual ways. Unlike the clear line of demarcation between the groups on the issue of abortion, some pro-choice advocates oppose all forms of human cloning and some pro-life advocates support so-called "therapeutic" cloning. It would seem that these unique alliances might allow more fruitful policy discussions. However, despite agreement across traditional boundaries on the substance of particular bills, some of the new alliances break down because the sponsors of the bills include preambles that evince either pro-life or pro-biotechnology sentiments that go far beyond the substance of the bills themselves.

This is a new development in the policy considerations related to human conceptuses. The legislators themselves are acting like lobbyists by including within their bills language espousing a particular position that goes far beyond the bills' actual legal substance. Some bills seem to be driven more by political posturing than by an attempt to actually get legislation enacted to deal with the issue at hand.

This paper surveys the state law landscape by analyzing 1) existing laws with implications for reproductive cloning, therapeutic cloning and embryo stem cell research, 2) proposed bills addressing those technologies, and 3) potential constitutional challenges to the laws.

The Existing Legal Framework

A. Existing Bans on Cloning

1. The Restrictions: There are eight states that have enacted explicit laws to ban reproductive cloning, and one state that prohibits only the use of state funds for reproductive cloning. Of these eight states, four also ban "therapeutic" cloning. Some states also ban shipping, transferring, or receiving for any purpose an embryo produced by human cloning. Under the existing reproductive cloning bans, some states ban the purchase or sale of an ovum, zygote, embryo, or fetus for the purpose of cloning a human being. Other states, however, have bans on payment for eggs, embryos, or human tissue that could also cover some of the transactions involved in reproductive or "therapeutic" cloning. California's initial anti-cloning bill, adopted in 1998, created an advisory committee that met frequently over a five year period, heard extensive testimony from scientists and members of the public, and issued a report.

2. *Exceptions for Certain Research:* In some states, the ban on reproductive cloning is accompanied by language supporting other types of research or medical practices. In Louisiana and Michigan, for example, the cloning bans say that they do not prohibit scientific research on a cell-based therapy. Similarly, some states allow the use of nuclear transfer or other cloning techniques to produce molecules, DNA, cells other than human embryos, tissues, organs, plants, or animals other than humans. The Arkansas and Rhode Island laws also specifically allow *in vitro* fertilization and fertility enhancing drugs, so long as they are not used in the context of human cloning. The Iowa law more broadly allows *in vitro* fertilization and the use of fertility drugs.

3. *Penalties:* The penalties for violation of existing cloning bans range widely. In Louisiana and Michigan, for example, penalties can be up to 10 years imprisonment and fines of $10 million for an entity such as a clinic or corporation and $5 million for an individual. In contrast, in Arkansas the penalty is a mere $250,000, but there are also felony criminal penalties. Moreover, in some states, cloning can result in the permanent revocation of a doctor's license and the denial of any other type of license or permit from the state regarding any trade, occupation or profession.

B. Existing Bans on Embryo Research

Twelve states' laws apply to *in vitro* embryos. In New

Hampshire, the regulation of research on embryos prior to implantation is minimal. The research must take place before day 14 post-conception, and the subject embryo must not be implanted in a woman. These stipulations could readily be met by researchers wanting to use IVF embryos as a source of stem cells.

Nine states ban research on *in vitro* embryos altogether, and two states ban destructive embryo research. In Louisiana, an *in vitro* fertilized ovum may not be farmed or cultured for research purposes; the sole purpose for which an IVF embryo may be used is for human *in utero* implantations.30 In the other states, embryo research is banned as part of the broader ban on all research involving live conceptuses. These laws ban embryo stem cell research. The penalties are high — in some states, the punishment includes imprisonment. In some instances, the woman or couple who donate an embryo or fetus for research purposes also face liability. Laws in Maine, Michigan, North Dakcta, and Rhode Island prohibit the transfer, distribution, or giving away of any live embryo for research purposes. In Maine, a person who does so is subject to a fine of up to $5,000 and up to five years' imprisonment.

C. Existing Laws Specifically Addressing Embryo Stem Cell Research

A California law declares that research shall be permitted on human embryonic or adult stem cells from any source, including somatic cell nuclear transfer. Such research must be reviewed by an approved institutional review board and such research may not be undertaken without written informed consent of the embryo donor. Interestingly, the law does not say consent of the "donors," plural. So it would appear that the female patient of infertility services (and not the woman and her husband) is the sole source of consent. California also enacted a law urging Congress to ban reproductive cloning, while permitting therapeutic cloning and embryo stem cell research.

II. Proposed Bills

A. Preambles to the Bills – Hype versus Doom

Legislators are currently considering bills that govern the technologies of cloning or embryo stem cell research. But even when the provisions of those bills are narrowly tailored to deal with those specific technologies, lawmakers include extensive preambles that include sweeping pro-life or pro-biotechnology language. None of the bills include a pro-choice preamble.

A Kentucky bill attempts to fuel a biotechnology sector in the state. Its preamble mentions the "great potential" of nuclear transplantation to treat "diseases and disorders, including but not limited to Lou Gehrig's disease, Parkinson's disease, Alzheimer's disease, spinal-cord injury, cancer cardiovascular diseases, diabetes, rheumatoid arthritis, and many others." Its provisions would require anyone engaging in therapeutic cloning to register with the health department and pay a $50 fee. The bill also implores public colleges in Kentucky to protect "the interest of the Commonwealth or the institution relating to all intellectual property and other rights to any research, experiments, or other activity related to human nuclear transplantation. . . ."

On the other hand, some bills' preambles are designed to cast aspersions on the technologies. The preamble of a Virginia Senate resolution contains a hodge-podge of historical facts (from slavery in Virginia 1619 to 1865 to Advanced Cell Technologies cloning of an embryo), Presidential statements (from President Reagan through President George W. Bush), and concerns about eugenics and commodification of humans. The resolution would establish a legislative task force, but its agenda is directed in a onesided way. It is supposed to "examine the medical, ethical and scientific policy implications of prohibiting the creation of embryos *in vitro* for any purpose other than bringing them to birth, and the criminalizing of the transfer of compensation, in cash or in-kind, to induce any person to donate sperm or eggs for any purpose other than procreation. The joint subcommittee shall also examine the efficacy of research using adult stem cells rather than embryonic stem cells."

Some bills assert scientific or medical facts that are not necessarily accurate. For example, Louisiana's resolution says, "stem cells derived from adult or placental tissues may hold greater promise than embryonic stem cells for human medicine, due to a lesser danger of immune reactivity against a recipient."

B. Proposed Bills on Reproductive Cloning

1. The Restrictions: In 2002 and 2003, 27 states introduced bills that would ban reproductive cloning. One bill details a specific enforcement mechanism for the human cloning ban. A New Jersey bill provides that the Commissioner of Health and Senior Services may issue a cease and desist order, with a subsequent administrative hearing within 20 days, with judicial review available.

Many of these proposed laws suffer from drafting infirmities. For example, some of the states' proposals create a loophole by only prohibiting the reproduction of a "genetically identical" individual through cloning. Since the current cloning technique uses a donated egg to create a clone, the resulting individual will have some additional mitochondrial DNA from the enucleated egg, so he or she will not be genetically identical to the original individual. In contrast, other states' bills prohibit the reproduction of a "virtually genetically identical" individual through human cloning using a human or non-human egg, which would apply even if a donated egg, containing additional mitochondria were used.

Certain states' bills would prohibit the transferring of a human cell into a human egg cell. But, in January 1998, scientists at the University of Wisconsin revealed that cow eggs could serve as incubators for nucleic DNA of other mammalian species. Thus, the proposed laws could be evaded by making a human clone via somatic cell DNA transfer into an enucleated cow egg. Other states' bills prohibit human reproductive cloning using a human or nonhuman egg.

A New Jersey bill has the opposite problem. Rather than being too narrow, it is too broad, using vague language in places and even potentially banning the creation of children through sexual intercourse. The focus of the bill is to ban human cloning and germline genetic engineering. But it defines human cloning as the replication of one or more human beings, through "sexual or asexual reproduction." When two people create a child sexually are they replicating themselves? The key word, "replication," is not defined.

The bill prohibits the destruction of human or non-human embryos "for the purpose of human cloning." There is a certain vagueness to that provision. If a researcher working on primate cloning generates data that might be useful to cloning human beings could he or she be prosecuted?

2. *Public funds:* Six states' bills would prohibit the use of public funds and facilities to participate in cloning or attempted cloning.

3. *Other Means to Discourage Cloning:* Some states' bills attempt to discourage reproductive cloning by prohibiting either the sale or purchase or the shipment or receival of an ovum, zygote, embryo, or fetus to clone a human. Some

states' laws attempt to discourage cloning through liability provisions. A Florida bill would create vast liabilities as another means to discourage cloning. Participants in human reproductive cloning would be liable to "the individual, the individual's spouse, dependents, and blood relatives, and to any woman impregnated with the individual, her spouse, and dependents, for damages for all physical, emotional, economic, or other injuries suffered by such persons at any time as a result of the use of human cloning to produce the individual." Even more dramatically, participants in human reproductive cloning would be liable to the cloned individual and his or her legal guardian for support until the age of majority (or longer, if the clone has "any congenital defects or other disability" related to the cloning).

4. Exceptions: Eighteen states' bills have an exception allowing cloning to create DNA, tissues, and organs. Eleven states' bills would not restrict the use of *in vitro* fertilization or fertility enhancing drugs.

C. Proposed Bills on Therapeutic Cloning

There are 26 states that have bills about therapeutic cloning. Two states have at least one bill that would ban therapeutic cloning and at least one that would allow therapeutic cloning.

1. Ban Therapeutic Cloning: Twenty-six states have bills that would prohibit all human cloning including therapeutic cloning. Some states prohibit therapeutic cloning through a general cloning ban that defines human cloning as the creation of any human organism through somatic cell nuclear transfer. Others prohibit therapeutic cloning by banning not only the creation of the cloned embryo, but also the "derivation of any product from human cloning."

2. Allow Therapeutic Cloning: Two states have bills

that specifically allow therapeutic cloning. In addition, twelve other states allow for the use of stem cells from any source, including those derived from somatic cell nuclear transplantation. One state bill would encourage researchers to do stem cell research using only human adult or placental tissues obtained with informed consent, and would discourage research involving human embryonic or fetal tissue. One state's bill urges state universities to refrain from embryonic stem cell research.

D. Proposed Bans on Embryo Stem Cell Research

1. Ban Embryo Stem Cell Research: Five states have bills that would prohibit acts where a human fetus or embryo is destroyed or subject to injury. New Jersey has these provisions in a cloning bill, while Wisconsin has them in an artificial insemination/*in vitro* fertilization bill. The others (Kansas, Michigan, and Virginia) have these provisions in stem cell research bills. One state's bill would make all contracts for payment for the extraction or use of embryonic stem cells against public policy, void, and unenforceable. In addition, a Michigan bill, in its preamble, opposes the White House decision to allow certain embryonic stem cell research. The substance of the bill is a provision urging the state's public universities to refrain from embryonic stem cell research.

2. Allow Embryo Stem Cell Research, Including Therapeutic Cloning: Twelve states have bills that would allow embryo stem cell research from any source. While the bills literally say "from any source," some also have a list saying this includes human embryonic stem cells, human embryonic germ cells, and human adult stem cells, from any source, including somatic cell nuclear transplantation. In the states where proposed bills would permit embryonic stem cell research, there are often provisions regarding how the research should be conducted. In eleven states, for example, physicians undertaking *in vitro* fertilization on patients must discuss options for the disposition of frozen embryos, including donation to research, and written informed consent for such donation would have to be obtained. In at least one state, the proposed research must be assessed by a state department, and in eleven states the research must be assessed by an institutional review board. A Kentucky bill requires that individuals engaged in therapeutic cloning must register their name, address, and phone number, which information shall be public record. Under one proposed California bill, the State Department of Health Services would have the responsibility to develop guidelines regarding the derivation or use of human embryonic stem cells.

In Illinois, care was taken in the proposed bill favoring embryonic stem cell research not to create barriers for the research due to contract or tort liability. Consequently, the proposed Illinois bill states that "procuring, furnishing, donating, processing, distributing, or using embryonic or cadaveric fetal tissue for research purposes pursuant to this Act is declared for the purposes of liability in tort or contract to be the rendition of a service by every person, firm, or corporation participating therein, whether or not remuneration is paid, and is declared not to be a sale of any such items and no warranties of any kind or description nor strict tort liability shall be applicable thereto."

3. Address Embryo Donation: Twelve states' bills would expressly allow a person to donate human embryos for research. Twelve states' bills would also prohibit the sale of embryonic or cadaveric fetal tissue for research purposes.

4. Review Committees: Six states' bills would establish a subcommittee or task force to review and/or study issues related to stem cell research. Some of these states have a bill that only establishes a task force. Some states establish this task force, subcommittee, or review board within a bill relating to stem cell research. Maryland has one of each. It has often been true in the bioethical area that when controversial new technologies are introduced, states propose commissions to address the issue. Often, the creation of the commission is a stalling tactic to avoid having to confront the issue head on or to postpone having to limit a technology, or to facilitate a moratorium on the technology (while the Commission evaluates the technology). Sometimes, though, state commissions have admirably overseen a technology's development, application and evaluation. Such was the case in the 1970's and 1980's with Maryland's newborn screening commission. So it is no surprise that of the bills proposing a state task force or commission, the Maryland bill provides the best guidance about what the commission would do. The Maryland task force would:

> "(1) Make recommendations for state policy regarding stem cells derived from embryos created solely for research purposes;
>
> (2) evaluate whether there should be written informed consent requirements for prospective donors of embryonic stem cells derived from embryos created during infertility treatment;
>
> (3) study constitutional, ethical, and policy issues with respect to whether the state should regulate or prohibit commerce in embryos or stem cells;
>
> (4) (i) determine what stem cell research is already being done in the state by private companies and academic institutions; and (ii) evaluate how these companies and institutions are regulating themselves when doing stem cell research;
>
> (5) determine the effect on Maryland businesses and institutions of Federal restrictions that limit federally

funded research and development to existing cell lines on stem cell research;

(6) analyze the roles of, and interrelationship between, federal and state oversight of stem cell research; and

(7) review any other matter relating to stem cell research that the task force considers necessary and proper."

5. Public Funding or Tax Credits for Stem Cell Research: States have also proposed bills allowing for or prohibiting the granting of state funds for stem cell research. California's proposed "Biomedical Research and Development Act of 2004" would award grants or loans to public or private institutions conducting research in stem cell biology. To be eligible for funding, all research projects must be reviewed and approved by an institutional review board.

The states of Indiana, Michigan, Mississippi, and Nebraska have gone the opposite direction and introduced bills specifically prohibiting the use of state money or tax credits for stem cell research. The Indiana bill proposes an amendment to the state tax code denying tax exemptions and credits for research involving human cloning or stem cell research with embryonic tissue. A Mississippi House Bill and a Mississippi Senate Bill provide that no public funds shall be used to assist in or to provide facilities for stem cell research that uses cells from human embryos and human cloning.

Finally, although not establishing public funds for embryonic stem cell research, six Florida bills seek to establish a center for "Universal Research to Eradicate Disease" which will facilitate funding opportunities for stem cell research.

E. Proposed Bills on Health Care Providers' Right of Conscience

In addition to proposing bills on the prohibition of or

allowance of embryonic stem cell research and cloning, a number of states have proposed bills protecting a health care provider, payer, or institution's right to refuse certain services based on moral or ethical convictions. Kansas, Louisiana, and Rhode Island have all introduced bills entitled "Health Care Providers Rights of Conscience."

The Kansas bill introduced on January 28, 2002, specifically allows all persons to refuse to participate in the provision of, or pay for, a health care service related to human cloning and embryonic stem cell and fetal experimentation. The Louisiana bill and the identical Rhode Island House and Senate bills allow all health care providers, institutions and payers to decline to counsel, advise, pay for, provide, perform, assist, or participate in providing or performing health care services that violate their consciences, including, human cloning and embryonic stem cell research.

F. Constitutionality of State Law

Not all regulations affecting research are constitutional. Laws restricting research on conceptuses may be struck down as too vague or as violating the right to privacy to make reproductive decisions. Such a challenge was successful in a federal district court case, Lifchez v. Hartigan, which held that a ban on research on conceptuses was unconstitutional because it was too vague in that it failed to define the terms "experimentation" and "therapeutic." The court pointed out that there are multiple meanings of the term "experimentation." It could mean pure research, with no direct benefit to the subject. It could mean a procedure that is not sufficiently tested so that the outcome is predictable, or a procedure that departs from present-day practice. It could mean a procedure performed by a practitioner or clinic for the first time. Or it could mean routine treatment on a new patient. Since the statute did not define the term, it was unconstitutionally vague. It violated researchers' and clinicians' due process rights under the fifth amendment since it forced them to guess whether their conduct was unlawful. A similar result was reached by a federal appellate court assessing the constitutionality of a Louisiana law prohibiting nontherapeutic experimentation on fetuses in Margaret S. v. Edwards. The appeals court declared the law unconstitutional because the term "experimentation" was so vague it did not give researchers adequate notice about what type of conduct was banned.

The court said that the term "experimentation" was impermissibly vague since physicians do not and cannot distinguish clearly between medical experimentation and medical tests. The court noted that "even medical treatment can be reasonably described as both a test and an experiment." This is the case, for example, "whenever the results of the treatment are observed, recorded, and introduced into the data base that one or more physicians use in seeking better therapeutic methods." A third case struck down as vague the Utah statute that provided that "live unborn children may not be used for experimentation, but when advisable, in the best medical judgment of the physician, may be tested for genetic defects."

The Tenth Circuit held "[b]ecause there are several competing and equally viable definitions, the term 'experimentation' does not place health care providers on adequate notice of the legality of their conduct." In a fourth case, Forbes v. Napolitano, a number of plaintiffs challenged the constitutionality of an Arizona statute that criminalized medical experimentation or investigation involving fetal tissue from induced abortions. The statute at issue in that litigation, A.R.S. § 36-2302(A), provides: A person shall not knowingly use any human fetus or embryo, living or dead, or any parts, organs, or fluids of any such fetus or embryo resulting from an induced abortion in any manner for any medical experimentation or scientific or medical investigation purposes except as is strictly necessary to diagnose a disease or condition in the mother of the fetus or embryo and only if the abortion was performed because of such disease or condition. An exception to this is found in A.R.S. § 36-2302(C), which reads: This section shall not prohibit any routine pathological examinations conducted by a medical examiner or hospital laboratory provided such pathological examination is not part of or in any way related to any medical or scientific experimentation. The penalty for violation of A.R.S. § 36-2302, a class 5 felony, is one and a half years in prison and fines up to $150,000. Plaintiffs argued that the statute prevented fetal tissue transplantation treatment of Parkinson's disease, certain fertility treatments, and development of treatments for illness.

The court ultimately affirmed the decision of the lower court in holding that the statute was unconstitutionally vague. Specifically, the "distinction between experiment and treatment in the use of fetal tissue is indeterminate, regardless of whether the tissue is obtained after an induced abortion." Furthermore, the court determined that a criminal statute that serves to prohibit medical experimentation but provides no guidance as to where to draw the line between experiment and treatment does not provide doctors with constructive notice, nor does it provide police, prosecutors, juries, and judges with standards for application. Specific bans on embryo stem cell research are unlikely to raise the same constitutional concerns. Part of the constitutional deficiency in the Arizona, Illinois, Louisiana, and Utah cases was that physicians offering health care services for their patients (such as embryo donation from an infertile woman, preimplantation genetic screening on an embryo, or treatment of a pregnant woman for diabetes) would not know if the activity would be considered by prosecutors to be experimental with respect to the embryo or fetus.

Moreover, the female patient's reproductive freedom was implicated statute. Stem cell researchers do not have those potential legal arguments. There is a slight possibility, however, a reproductive liberty challenge could be raised against a law banning reproductive cloning. Bills on cloning and stem cell research are the subjects of vast media attention and public debate. As technologies develop that focus increasingly on the human embryo, lawmakers are having difficulty coming to agreement on legislative policy in the absence of societal accord about the moral and legal status of the embryo.

COMMISSION ON GENETIC RESOURCES FOR FOOD AND AGRICULTURE

Tenth Regular Session Rome, 8 – 12 November 2004

Status of the Inter-Governmental Technical Working Group on Plant Genetic Resources for Food and Agriculture

Article 1 - Terms of Reference

The Intergovernmental Technical Working Group on Plant Genetic Resources for Food and Agriculture (the Working Group) shall:

- review the situation and issues related to agrobiodiversity in the area of plant genetic resources for food and agriculture and advise and make recommendations to the Commission on these matters;
- consider the progress made in implementing the Commission's programme of work on plant genetic

resources for food and agriculture as well as any other matters referred to the Working Group by the Commission;

- report to the Commission on its activities. In order for the Working Group to carry out this mandate, the Commission will assign specific tasks to the Working Group.

Article II - Composition

The Working Group shall be composed of twenty-seven Member Nations from the following regions:

5 from Africa

5 from Europe

5 from Asia

5 from Latin America and the Caribbean

3 from the Near East

2 from North America

2 from Southwest Pacific.

Article III - Election and term of office of Members

The Members of the Working Group shall be elected at each regular session of the Commission and serve until the next regular session of the Commission. They will be eligible for re-election.

Article IV - Officers

1. The Working Group shall elect its Chairperson and one or more Vice-Chairpersons from among the representatives of Members of the Working Group at the beginning of each session. These officers shall remain in office until the next session of the Working Group and shall be eligible for re-election.

2. The Chairperson, or a Vice-Chairperson in the absence

of a Chairperson, shall preside over the meetings of the Working Group and exercise such other functions as may be required to facilitate its work.

Article V - Sessions

The Commission shall decide on the timing and duration of the sessions of the Working Group, when required. In any case, the Working Group shall hold no more than one regular session annually.

Article VI - Observers

1. Members of the Commission which are not Members of the Working Group may participate, upon request to the Commission Secretariat, in the work of the Working Group in an observer capacity. 2. The Working Group, or the bureau on behalf of the Working Group, may invite experts, as well as representatives of specialized international organizations, to attend its meetings.

United States Regulatory Oversight in Biotechnology Responsible Agencies: Overview

The Agencies primarily responsible for regulating biotechnology in the United States are the US Department of Agriculture (USDA), Environmental Protection Agency (EPA), and the Food and Drug Administration (FDA). Products are regulated according to their intended use, with some products being regulated under more than one agency.

Agency	*Products Regulated*
U.S. Department of Agriculture	plant pests, plants, veterinary biologics
Environmental Protection Agency	microbial/plant pesticides, new uses of existing pesticides, novel microorganisms

Food and Drug Administration	food, feed, food additives, veterinary drugs,human drugs and medical devices

Before commercialization, genetically engineered plants/ organisms must conform with standards set by State and Federal marketing statutes such as State seed certification laws, the Federal Food, Drug, and Cosmetic Act (FFDCA), the Federal Insecticide, Fungicide, and Rodenticide Act (FIFRA), the Toxic Substances Control Act (TSCA), and the Federal Plant Pest Act. There are no national requirements for varietal registration of new crops.

EXAMPLES:

New Trait / Organism	*Regulatory Review Conducted by:*	*Reviewed for:*
Viral Resistance in food crop	USDA EPA FDA	Safe to grow Safe for the environment Safe to eat
Herbicide Tolerance in food crop	USDA EPA FDA	Safe to grow New use of companion herbicide Safe to eat
Herbicide Tolerancein ornamental crop	USDA EPA	Safe to grow New use of companion herbicide
Modified Oil con-tentin food crop	USDA FDA	Safe to grow Safe to eat
Modified flower color ornamental crop	USDA	Safe to grow
Modified soil bacteria degrades pollutants	EPA	Safe for the environment

U.S. Department of Agriculture - USDA

Within USDA, the Animal and Plant Health Inspection Service (APHIS) is responsible for protecting US agriculture from pests and diseases. Under the authority of the Federal Plant Pest Act, APHIS regulations provide procedures for obtaining a permit or for providing notification, prior to "introducing" a regulated article in the United States.

Regulated articles are considered to be organisms and products altered or produced through genetic engineering that are plant pests of that there is reason to believe are plant pests. The act of introducing includes any movement into (import) or through (interstate) the United States, or release into the environment outside an area of physical confinement.

The regulations also provide for a petition process for the determination of nonregulated status. Once a determination of nonregulated status has been made, the product (and its offspring) no longer requires APHIS review for movement or release in the US.

Environmental Protection Agency - EPA

EPA ensures the safety of pesticides, both chemical and those that are produced biologically. The BioPesticides and Pollution Prevention Division of the Office of Pesticide Programs (OPP) uses the authority of the Federal Insecticide, Fungicide, and Rodenticide Act (FIFRA) to regulate the distribution, sale, use and testing of plants and microbes producing pesticidal substances. Under the Federal Food, Drug and Cosmetic Act (FFDCA) EPA sets tolerance limits for substances used as pesticides on and in food and feed, or establishes an exemption from the requirement of a tolerance.

EPA also establishes tolerances for residues of herbicides used on novel herbicide-tolerant crops.

Under the authority of the Toxic Substances Control Act (TSCA), EPA's TSCA Biotechnology Program regulates microorganisms intended for commercial use that contain or express new combinations of traits. This includes "intergeneric microorganisms" formed by deliberate combinations of genetic material from different taxonomic genera.

Food and Drug Administration - FDA

As a part of the Department of Health and Human Services, FDA regulates foods and feed derived from new plant varieties under the authority of the Federal Food, Drug, and Cosmetic Act. FDA policy is based on existing food law, and requires that genetically engineered foods meet the same rigorous safety standards as is required of all other foods.

FDA's biotechnology policy treats substances intentionally added to food through genetic engineering as food additives if they are significantly different in structure, function, or amount than substances currently found in food. Many of the food crops currently being developed using biotechnology do not contain substances that are significantly different from those already in the diet and thus do not require pre-market approval. Consistent with its 1992 policy, FDA expects developers to consult with the agency on safety and regulatory questions.

RELEVANT LAWS - REGULATIONS - RULES

USDA: LAWS

- Federal Plant Pest Act - 7 USC 7B

REGULATIONS

- 7 CFR 340 - Introduction of Organisms and Products Altered or Produced Through Genetic Engineering Which are Plant Pests or Which There is Reason to Believe are Plant Pests
- Simplification of Requirements and Procedures for Genetically Engineered Organisms and Products

  Federal Register: May 2, 1997 (Volume 62, Number 85)
- Genetically Engineered Organisms and Products;

Notification Procedures for the Introduction of Certain Regulated Articles; and Petition for Nonregulated Status

Federal Register: March 31, 1993 (Volume 58, Number 60)

- Introduction of Organisms and Products Altered or Produced Through Genetic Engineering Which are Plant Pests or Which There is Reason to Believe are Plant Pests

  Federal Register: June 16, 1987 (Volume 52, Number 115)

EPA: LAWS

- Federal Insecticide, Fungicide, and Rodenticide Act (FIFRA) - 7 USC 136
- Federal Food, Drug, and Cosmetic Act (FFDCA) - 21 USC 9
- Toxic Substances Control Act (TSCA) - 15 USC 53

REGULATIONS

Regulations Under FIFRA/FFDCA

40 CFR Parts 152, 172, 174, 180

- Plant-Pesticides, Supplemental Notice; Proposed Rule

  Federal Register: May 16, 1997 (Volume 62, Number 95)
- Plant-Pesticides Subject to the Federal Insecticide, Fungicide, and Rodenticide Act; Proposed Rule

  Federal Register: November 23, 1994 (Volume 59, Number 225)
- Microbial Pesticides: Experimental Use Permits and Notifications

  Federal Register: September 1, 1994 (Volume 59, Number 169)

- Statement of Policy; Microbial Products Subject to the Federal Insecticide, Fungicide, and Rodenticide Act and the Toxic Substances Control Act

  Federal Register: June 26, 1986 (Volume 51, Number 123)

Regulations Under TSCA

40 CFR Part 725 - Reporting Requirements and Review Processes for Microorganisms

- Microbial Products of Biotechnology; Final Regulation Under the Toxic Substances Control Act

  Federal Register: April 11, 1997 (Volume 62, Number 70)

FDA: LAWS

- Federal Food, Drug, and Cosmetic Act (FFDCA) - 21 USC 9

REGULATIONS

- Statement of Policy: Foods Derived from New Plant VarietiesFederal Register: May 29, 1992 (Volume 57, Number 104).

DOCUMENTS OF THE TENTH REGULAR SESSION OF THE CGRFA

Rome, 8 - 12 November 2004

CGRFA-10/04/REPORT

Report of the Tenth Regular Session of the Commission on Genetic Resources for Food and Agriculture

Working Document

CGRFA-10/04/1

Draft Provisional Agenda

CGRFA-10/04/2

Provisional Annotated Agenda and Time-table

CGRFA-10/04/3

Overview of the FAO *Global System for the Conservation and Sustainable Utilization of Plant Genetic Resources for Food and Agriculture*, and its potential contribution to the implementation of the International Treaty on Plant Genetic Resources for Food and Agriculture

CGRFA-10/04/4

Report of the Second Session of the Intergovernmental Technical Working Group on Plant Genetic Resources for Food and Agriculture

CGRFA-10/04/5

Follow-up to the recommendations of the Commission on Genetic Resources for Food and Agriculture and its Working Group, regarding Plant Genetic Resources for Food and Agriculture

CGRFA-10/04/5 Add.1

Progress in the preparation of the Second *Report on the State of the World's Plant Genetic Resources for Food and Agriculture*

CGRFA-10/04/6

Report on the International Network of *Ex Situ* Collections under the Auspices of FAO

CGRFA-10/04/7

Progress in the Development of the Global Strategy for the Management of Farm Animal Genetic Resources

CGRFA-10/04/8

Report of the Third Session of the Intergovernmental

Technical Working Group on Animal Genetic Resources for Food and Agriculture

CGRFA-10/04/9

Progress Report on the process of the preparation of the *First Report on the State of the World's Animal Genetic Resources* and the *Report on Strategic Priorities for Action*

CGRFA-10/04/9 Add.1

Initiation of the Follow-up Mechanism for the implementation of Strategic Priorities for Action - Animal Genetic Resources

CGRFA-10/04/9 Add.2

Draft Outline for the *First Report on the State of the World's Animal Genetic Resources* including the *Report on Strategic Priorities for Action*

CGRFA-10/04/9 Add.3

Budgetary and other aspects of a possible International Technical Conference on animal genetic resources

CGRFA-10/04/10.1

Report from FAO on its policies, programmes and activities on agricultural biological diversity:

(1) Sectorial matters

CGRFA-10/04/10.2

Report from FAO on its policies, programmes and activities on agricultural biological diversity:

(2) Cross-sectorial matters

CGRFA-10/04/10.3

Report from FAO on its policies, programmes and activities on agricultural biological diversity:

(3) Priority Areas for Inter-disciplinary Action (PAIAs)

CGRFA-10/04/11.1

Reports from international organizations on their policies, programmes and activities on agricultural biological diversity:

(1) United Nations and other Inter-governmental Organizations

CGRFA-10/04/11.2

Reports from international organizations on their policies, programmes and activities on agricultural biological diversity:

(2) International Agricultural Research Centres of the Consultative Group on International Agricultural Research (CGIAR)

CGRFA-10/04/11.3

Reports from international organizations on their policies, programmes and activities on agricultural biological diversity:

(3) International Non-governmental Organizations

CGRFA-10/04/11 Add.1 Rev.1

Reports from international organizations on their policies, programmes and activities on agricultural biological diversity:

Reports arrived late for translation

CGRFA-10/04/12

Cooperation with the Convention on Biological Diversity

CGRFA-10/04/13

Progress in the Draft Code of Conduct on Biotechnology as it relates to genetic resources for food and agriculture: policy issues, gaps and duplications

CGRFA-10/04/14

Future work of the Commission

Information Document

CGRFA-10/04/Inf.1 Statutes of the Commission on Genetic Resources for Food and Agriculture

CGRFA-10/04/Inf.2

Statutes of the Intergovernmental Technical Working Group on Animal Genetic Resources for Food and Agriculture, and Members elected by the Ninth Regular Session of the Commission

CGRFA-10/04/Inf.3

Statutes of the Intergovernmental Technical Working Group on Plant Genetic Resources for Food and Agriculture, and Members elected by the Ninth Regular Session of the Commission

CGRFA-10/04/Inf.4

Report of the Meeting on the Evaluation of the Pilot Phase of the New Approach for Monitoring the Implementation of the Global Plan of Action for the Conservation and Sustainable Utilization of Plant Genetic Resources for Food and Agriculture.

CGRFA-10/04/Inf.5

Indicators and reporting format for monitoring the implementation of the Global Plan of Action for the Conservation and Sustainable Utilization of Plant Genetic Resources for Food and Agriculture

CGRFA-10/04/Inf.6

Country Progress Report on the implementation of the *Global Plan of Action for the Conservation and Sustainable Utilization of Plant Genetic Resources for Food and Agriculture*

CGRFA-10/04/Inf.7

Strengthening Seed Systems

CGRFA-10/04/Inf.8

Country Guidelines for the Second *State of the World's Plant Genetic Resources for Food and Agriculture*

CGRFA-10/04/Inf.9

Draft Report on the Strategic Priorities for Action

CGRFA-10/04/Inf.10

Statement of competence and voting rights submitted by the European Community and its Member States

CGRFA-10/04/Inf.11

Agreement Between the Tropical Agricultural Research and Higher Education Center (CATIE) and the Food and Agriculture Organization of the United nations (FAO) Placing Collections of Plant Germplasm under the Auspices of FAO

CGRFA-10/04/Inf.13

Reports from International Organizations on their Policies, Programmes and Activities on Agricultural Biological Diversity Part II: The International Agricultural Research Centres of the Consultative Group on International Agricultural Research (CGIAR) - List of National and International Partners in Research Programmes on Farm Animal Genetic Resources

CGRFA-10/04/Inf.14

Report on the International Network of Ex Situ Collections under the Auspices of FAO: Further Information Provided by the International Centre for Tropical Agriculture (CIAT), Regarding its Request for a Re-examination of U.S. Patent No.5,894,079

CGRFA-10/04/Inf.16

Strengthening National Structures for the Management of Farm Animal Genetic Resources - Results of a Questionnaire Survey

CGRFA-10/04/Inf.17

List of documents

CGRFA-10/04/Inf.18

List of Delegates and Observers

CGRFA-10/04/Inf.19

Memorandum of understanding between the Food and Agriculture Organization of the United Nations (FAO) and the International Plant Genetic Resources Institute (IPGRI)

COMMISSION ON GENETIC RESOURCES FOR FOOD AND AGRICULTURE

Tenth Regular Session Rome, 8 – 12 November 2004

Statutes of the Commission on Genetic Resources for Food and Agriculture

1. Membership

The Commission shall be open to all members and associate members of the Organization. It shall be composed of those members or associate members, which notify the Director-General of their desire to be considered as members of the Commission.

2. Terms of Reference of the Commission

The Commission shall have a coordinating role and shall deal with policy, sectoral and cross-sectoral matters related to the conservation and sustainable use of genetic resources of relevance to food and agriculture. Its terms of reference shall be:

i) to keep under continuous review all matters relating

to the policy, programmes and activities of FAO in the area of genetic resources of relevance to food and agriculture, including their conservation and sustainable use and the fair and equitable sharing of benefits derived from their utilization, and to advise the Director-General and the Council and, as appropriate, its technical committees, including in particular the Committees on Agriculture, Forestry and Fisheries, on such matters;

ii) to recommend such measures as may be necessary or desirable to ensure the development, as appropriate, of a comprehensive global system or systems on genetic resources of relevance to food and agriculture and to monitor the operation of its/their components, in harmony, where applicable, with the Convention on Biological Diversity and other relevant international instruments;

iii) to provide an intergovernmental forum for negotiations and to oversee the development, upon the request of the FAO Governing Bodies, of other international agreements, undertakings, codes of conduct or other instruments relating to genetic resources of relevance to food and agriculture, and to monitor the operation of such instruments;

iv) to facilitate and oversee cooperation between FAO and other international governmental and non-governmental bodies dealing with the conservation and sustainable use of genetic resources, in particular with the Conference of Parties to the Convention on Biological Diversity and the UN Commission on Sustainable Development, and to seek to develop appropriate mechanisms for cooperation and coordination in consultation with such bodies;

v) subject to approval by the Governing Bodies of FAO,

as appropriate, to respond to requests from the Conference of the Parties to the Convention on Biological Diversity in the specific area of genetic resources of relevance to food and agriculture, including the provision of information and other services to the Conference of the Parties and its subsidiary bodies, especially in the areas of early warning systems, global assessment and clearing house facilities, in particular, and as appropriate, through the Global System for the Conservation and Utilization of Plant Genetic Resources for Food and Agriculture.

3. Intergovernmental Technical Sectoral Working Groups of the Commission

i) The Commission may establish intergovernmental technical sectoral working groups ("Sectoral Working Groups"), with appropriate geographical balance, to assist it in the areas of plant, animal, forestry and fisheries genetic resources.

ii) The purpose of the Sectoral Working Groups shall be to review the situation and issues related to agrobiodiversity in the areas under their respective competences, to advise and make recommendations to the Commission on these matters and to consider the progress made in implementing the Commission's programme of work, as well as any other matters referred to them by the Commission.

iii) The composition and terms of reference for each Sectoral Working Group shall be established by the Commission.

4. Sessions of the Commission and its Sectoral Working Groups

i) The Commission shall normally hold one regular

session each biennium. It may also decide to convene extraordinary sessions as necessary, subject to the approval of the FAO Council. Sessions of the Commission shall normally be held at the Organization's Headquarters.

ii) The Sectoral Working Groups, where established, shall hold no more than one regular session annually.

5. Other Subsidiary Bodies

The Commission may establish such other Subsidiary Bodies as it may deem necessary for the effective discharge of its functions.

6. Funding of Sectoral Working Groups and other Subsidiary Bodies

i) The establishment of any Sectoral Working Group or other Subsidiary Body shall be subject to the determination by the Director-General that the necessary funds are available in the relevant chapter of the budget of the Organization or from extra-budgetary sources.

ii) Before taking any decision involving expenditure in connection with the establishment of Subsidiary Bodies, the Commission shall have before it a report from the Director-General on the programme, administrative and financial implications thereof.

7. Reporting

The Commission shall report to the Director-General, who shall bring to the attention of the Conference through the Council any recommendations adopted by the Commission which have policy implications or which affect the programme or finances of the Organization. As soon as they become available, copies of each report of the Commission will be circulated to members and associate members of the

Organization and also to international organizations and agencies that are concerned with genetic resources.

8. Secretariat and expenses

i) The Secretary of the Commission shall be appointed by the Director-General and shall be administratively responsible to him. The Secretariat of the Commission will monitor and coordinate the preparations for the Commission meetings and the work for the Sectoral Working Groups, where established. The expenses of the Secretariat of the Commission shall be determined and paid by the Organization within the limits of the relevant appropriations in the approved budget of the Organization.

ii) Secretariat services for each Sectoral Working Group, where established, will be provided by the respective technical division of FAO as part of its annual programme of work.

iii) Expenses incurred by representatives of members of the Commission and its Working Groups, their alternates and advisers, when attending sessions of the Commission, its Sectoral Working Groups or other Subsidiary Bodies, as well as the expenses of observers at sessions, shall be borne by the respective governments or organizations.

9. Observers

Meetings of the Commission shall be open to observers from members and associate members that are not members of the Commission, from states that are nont members or associate members of the Organization, and from international organizations, in accordance with the relevant provisions of the rules and principles adopted by the Conference.

10. Rules of procedure

The Commission may adopt and amend its own rules of procedure, which shall be in conformity with the Constitution and the General Rules of the Organization and with the statement of principles governing commissions and committees adopted by the Conference.

Recent Developments in Stem Cell Research and Therapy

Research using human and animal stem cells is an extremely active area of current biomedical inquiry. It is contributing new knowledge about the pathways of normal and abnormal cell differentiation and organismal development. It is opening vistas of new cell transplantation therapies for human diseases.

Although the availability of a variety of human stem cells is relatively recent—the isolation of human embryonic stem cells was first reported only in 1998—much is happening in both publicly funded and privately funded research centers around the world. It is difficult for anyone to stay abreast of all the results now rapidly accumulating.

To help us fulfill our mandate to "monitor stem cell research," the President's Council on Bioethics asked several experts to survey the recent published scientific literature and to contribute articles on various areas of stem cell research to this report (see articles by Drs. Gearhart, Ludwig and Thomson, Verfaillie, Prentice, Itescu, and Jaenisch in the Appendices). These reviews and the present chapter emphasize peer-reviewed, published work with human stem cells through July 2003. Interested readers should also consult the wide variety of other review articles that have appeared. This chapter should be read in conjunction with the commissioned review articles cited above. It draws on their findings, as well as on the Council's own monitoring activities, but it makes no attempt to summarize all the complexity of stem cell research or the vast array of results. Rather we offer here some general observations and specific examples that might help non-scientist readers understand the overall state of present human stem cell research, its therapeutic promise, and some of the problems that need to be solved if the research is to yield sound knowledge and clinical benefit. To that end, we highlight the importance of well-characterized, stable preparations of stem cells for obtaining reproducible experimental results, and we identify several problems that must be solved before these requirements can be fully met.

This study describes, by way of illustration and example, some of the better-characterized adult and embryonic stem cells. It also indicates some of the specific investigations that are being conducted with their aid. Finally, it considers how human stem cells are being used to explore their potential for treating disease, using experiments in animal models of Type- 1 diabetes as an example, and it points out some of the difficulties that must be overcome before stem cell-based remedies may be available to treat human diseases.

We confine our attention here to newly identified types of human stem cells and their potential use in research and future medical treatment. Accordingly, we do not consider those stem cell types that are already well established in medical practice and research. Specifically, we will not examine those preparations of bone marrow cells that have been clinically used for some years to treat various forms of anemia and cancer.

Neither will we deal with hematopoietic (bloodforming) stem cells that have been isolated and purified from bone marrow and are now being intensively studied. Although these developments lie beyond the scope of this report, the demonstrated usefulness of these cells for research and therapy encourages many researchers to expect similar benefits from the newer stem cells that we shall consider here.

Cells and Their Derivatives

The adult human body, and all its differentiated cells, tissues, and organs, arise from a small group of cells contained within the early embryo at the blastocyst stage of its development. During in vivo embryonic development, these cells, constituting the inner cell mass (ICM), will divide and differentiate in concert with each other and with the whole of which they are a part, eventually producing the specialized and integrated tissues and organs of the body. But when embryos are grown [using in vitro fertilization (IVF)] in a laboratory setting, these ICM cells may be removed and isolated, and under appropriate conditions some will proliferate in vitro and become embryonic stem cell lines.

These embryonic stem cells are capable of becoming many different types of differentiated cells if stimulated to do so in vitro. However, it is not yet clear that the cells that survive the in vitro selection process to become embryonic stem cells have all of the same biological properties and potentials *as t*he ICM cells of the blastocyst. In particular, it is not known for certain that human embryonic stem cells in vitro can give rise to all the different cell types of the adult body. As no*ted* in the Introduction to this report, stem cells are a diverse class of cells, which can now be isolated from a variety of embryonic, fetal, and adult tissues. Stem cells share two characteristic properties: (1) unlimited or prolonged selfrenewal (that is, the capacity to maintain a pool *of stem cel*ls like themselves), and (2) potency for differentiation, the potential to produce *more differentiated cell t*ypes—usually more than one *and, in some cases,* many.

When stem cells head down the pathway toward differentiation, they usually proceed by first giving rise to a more specialized kind of stem cell (sometimes called "precursor cells" or "progenitor cells"), which can in turn either proliferate through self-renewal or produce fully specialized or differentiated cells.

At the top of the figure is an undifferentiated stem cell; in the central box are more "specialized" stem cells (or "precursor cells" or "progenitor cells"); at the bottom are various differentiated cells that are derived from the specialized stem cells. Dashed arrows indicate symmetrical (in the sense that both the daughter cells are stem cells) cell divisions that produce more stem cells (self-renewal). Solid arrows indicate asymmetric cell divisions that produce more differentiated daughter cells. (There may also be self-renewal with asymmetric division—not shown here—in which one daughter cell initiates a differentiation pathway while the other remains a stem cell.) Differentiation signals can be supplied by both soluble proteins and by specific, cell-surface binding sites.

Some of the specialized stem cells inside the dashed box, for example, mesenchymal stem cells, can be isolated from tissues after birth and correspond to adult stem cells. Scientists are currently investigat*ing wh*ether, at least in some cases, the process can be reversed, that is, whether specialized cells may, on appropriate signals, dedifferentiate to become precursor or even fully und*ifferentiated st*em cells.

The terminology used to describe different stem cell types can be confusing. As used in this chapter, stem cells are selfrenewing, cultured cells, grown an*d preserved* in vitro, that are capable—upon exposure to appropriate signals—of differentiating themselves into (usually more than one) specialized cell types. Stem cells may be classified either according to their origins or according to their developmental potential.

Stem cells may be obtained *from various sources: f*rom embryos, from fetal tissues, from umbilical cord blood, and from tissues of adults (or children). Thus, depending on their origin, stem cell preparations may be called adult st*em cel*ls, embryonic stem cells, embryonic germ cells, or fetal stem cells.

Adult stem cells are cells derived from various tissu*es or organs in h*umans or animals that have the two characteristic properties of stem cells (self-renewal and potency for differentiation). Embryonic stem cells (ESCs) are derived from cells is*olated from the inner* cell mass of early embryos. Embryonic germ cells (EGCs) are stem cells. Depending on their developmental potential, cells may be called pluripot*ent, multipotent, or un*ipotent. Cells that can produce all the cell types of the developing body, such as th*e IC*M cells of the blastocyst, are said to be pluripotent. The somewhat more specialized stem cells*, of the so*rt found in the developed organs or tissues of the body, are said to be multipotent if they produce more than one differentia*ted tissue c*ell type, and unipotent if they produce only one differentiated tis*sue cell t*ype. We introduce in this chapter an additional term: stem cell preparation. A stem cell preparation is a p*opulation of stem cel*ls, *prepared, grown, and p*reserved under certain conditions. Because different laboratories (or even the same one) can have different preparations of the same type of stem cell, it is important to recognize the potential differences between particular preparations of embryonic stem cells. It will sometimes be important to call attention to this fact, by speaking of a "preparation of ES cells" (or a preparation of adult s*tem cells) r*ather than of "ES cells," pure and simple.

We will use the term "stem cell preparations" when we are speaking of a diverse group of stem cell cultures, when we are speaking of stem cell cultures that contain an admixture of other types of cells, or when the developmental homogeneity of the stem cells in the population has not been defined.

Adult and embryonic stem cell populations have also been called "stem cell lines." In the past, the term "cell line" denoted a cell population (usually of cancer cells containing abnormal chromosome numbers or structure, or both) that could grow "indefinitely" in vitro. Embryonic and some adult stem cell preparations are capable of prolonged growth beyond 50 population doublings in vitro while retaining their characteristic stem cell properties and initially with no change in the chromosome numbers and structure. It is not yet known whether any preparation of human ES cells (generally believed to be much longer-lived than adult stem cells) will continue to grow "indefinitely," without undergoing genetic changes.

Under the influence of various cell-differentiation signals, embryonic stem cells differentiate into numerous distinct types of more specialized cells. Some of these are specialized stem cells that can also self-renew, while retaining their ability also to differentiate into multiple cell types. Recent research has led to the isolation of an increasing number of adult (non-embryonic) stem cells (dashed box area of Figure 1) from such tissues as bone marrow (for example, hematopoietic and mesenchymal stem cells), brain (for example, neural stem cells) and other tissues.

Although these stem cell preparations differ from one another in their future fates, they tend to be grouped together (especially in the public policy debates) under the name "adult stem cells," even though they may have been obtained from children or even from umbilical cord blood obtained at the time of childbirth.

Subsequent exposure to additional differentiation signals can cause these specialized stem cells to differentiate further, so that they finally give rise to the variety of differentiated cells that make up the adult body (labeled A-D in Figure 1). At each stage of the differentiation process, specific sets of genes are expressed (or "turned on") and other sets are repressed (or "turned off"), to produce the specific proteins that give each cell its distinctive properties. At each stage along the way, proteins called transcription factors play key roles in determining which sets of genes are expressed and repressed, and therefore what sort of a cell the newly differentiated cell will become.

Reproducible Results Using Stem Cell Preparations and THeir Derivatives

A major goal of scientific research is the acquisition of reliable knowledge based on experiments that yield reproducible results. Reproducible results are possible only if the materials used in experiments remain constant and stable.

To obtain reproducible results in experiments using stem cells, it is essential to produce, preserve, characterize, and continually re-characterize preparations of stem cells in ways that increase the likelihood that the cells used to repeat experiments will remain unchanged—a technically challenging task. The tendency of stem cells in vitro to differentiate spontaneously into more specialized cells makes the task of obtaining homogeneous and stable stem cell preparations especially challenging, and much basic research is needed to learn how to control the fate of these cells. Failure to control the cells may yield experimental results that are difficult or impossible to reproduce. The following more specific observations make clear the dimensions of this difficulty.

A. Initial Stem Cell Preparations: Can Contain Multiple Cell Types Isolation of adult stem cells from source tissues such as bone marrow, brain, or muscle initially yields a heterogeneous cell preparation. The initial preparation contains the several cell types found in the source tissue, and it may also include red blood cells, white blood cells, and (possibly) circulating stem cells, owing to the presence of blood flowing through the tissue in question. Initial mixtures of cells may then be treated in various ways to remove unwanted contaminating cells, thereby increasing the proportion of stem cells in the preparation. But seldom, if ever, does one produce an adult stem cell preparation that is 100 percent stem cells, unless the adult stem cell preparation has been "single-cell cloned" in vitro.

The way in which human embryonic stem cells have been produced from ICM cells also raises a question about the "species homogeneity" of the initial cell preparations. In the past, human embryonic stem cells were isolated and maintained by *in vitro* growth on top of irradiated (so that they no longer divide) "feeder layers" of mouse cells. It is thought that the feeder cells secrete factor(s) that enable the stem cells to divide while maintaining a relatively undifferentiated state.

Although the mouse cells have been treated to prevent their cell division, should any of them happen to survive, human embryonic stem cells prepared in this way may contain some viable mouse cells. More recently, several groups have shown that it is possible to grow ESCs on feeder layers of human cells, including fibroblasts obtained from skin biopsies, or without any feeder cell layer at all. One way to be certain that human embryonic stem cell preparations do not contain any mouse feeder cells is through "single cell cloning".

B. Genetically Homogenous Stem Cells through Single Cell Cloning: Some preparations of stem cells growing in vitro have been "single cell cloned," that is, grown as a population derived from a single stem cell. By placing a cylinder over a single *cell l*ocated with a microscope, scientists are able to isolate within the cylinder all the progeny produced by subsequent cell divisions beginning from this single cell. The result is a stem cell preparation in which all the cells are descended from the original single cell. The cells within the cylinder are then harvested and grown to greater numbers in vitro, and the resulting stem cell preparation is said to be "single cell cloned." The stem cells within a "single cell cloned" population are, at least to begin with, genetically homogeneous because they are all derived *from the sam*e original cell. Some of the ESC preparations produced prior to August 9, 2001 have been "single cell cloned." * The issue of possible mouse virus contamination is dealt with in Section F, below.

C. Expansion in Vitro, Preservation, and Storage : Reproducible results require that preparations of stem cells, even if genetically homogenous when first isolated, remain stable over time and during preservation. This, too, is not a simple matter with stem cells, despite the fact that the selfrenewal characteristic of human embryonic and adult stem cells enables them—unlike differentiated cells from many human tissues—to be grown in large numbers in vitro while maintaining their essential stem cell characteristics. After such expansion, many, presumably identical, vials of the cells can be frozen and preserved at very low temperatures. Frozen stem cell preparations can later be thawed and grown again in vitro to produce larger numbers of cells.

As with all dividing cells, stem cells are subject to a very small but definite chance of mutation during DNA replication; thus, prolonged growth in vitro could introduce genetic *heterogeneity into an originally homogeneous population. During this process of repeated expansion and prese*rvation, subtle changes in the growth conditions or other variables may give rise to "selective pressures" that can increase the heterogeneity in a stem cell preparation by favoring the multiplication of advantaged cell variants in the population. It is not known at present how many of the 78 human ESC preparations, designated as eligible for federal funding under the current policy, have developed genetic variants that may make them unsuitable for further research.

Whether several cycles of freezing and thawing change the phenotypic characteristics of stem cell preparations needs detailed study. However, the practical advantages of preserving stem cell preparations by freezing are too large to ignore. Such preservation makes it possible to repeat an experiment many times with a very similar stem cell preparation. It would also make it possible, should stem cell based therapies be developed in the future, to treat multiple patients with a common, well-characterized cell preparation derived from a single initial stem cell sample.

D: Chromosome Changes: In addition to the possible loss of homogeneity in stem cell preparations owing to variability in growth conditions or to freezing and thawing, there is the possibility of variation being introduced during the processes of growth and cell division.

Normal human stem cells (like all human somatic cells) have 46 chromosomes. During the copying of chromosomal DNA and the separation of daughter chromosomes at cell division, rare mistakes occur that lead to the formation of abnormal chromosomes or maldistribution of normal ones. Cells with abnormal chromosomes or chromosome numbers can progress to malignancy, so retention of the normal human chromosome number and structure is an essential characteristic of useful human stem cell preparations. The most studied preparations of human stem cells generally have normal human chromosome numbers and structure. Nevertheless, vigilance is needed, for even a small number of chromosomally abnormal cells could end up causing cancer in future clinical trials of stem cell based therapies.

E. Developmental Heterogeneity of Stem Cell Preparations: The in vitro growth conditions and the presence of specific chemicals or proteins, or both, in the culture medium can influence the differentiation pathway taken by stem cells as they start to differentiate. Thus, even initially homogeneous, "single cell cloned" stem cell preparations may become developmentally heterogeneous over time, with respect to the percentage of cells in the preparation that are in one or another differentiated state. For example, a stem cell preparation after growth in vitro under specific conditions might contain 75 percent fully differentiated (insulin-producing) cells and 25 percent partially differentiated cells. The biological properties of the fully differentiated cells and the partially differentiated cells are likely to be different. If such a cell preparation is used in research, or transplanted into an animal model of human disease and a biological effect is observed, one must do additional experiments to determine whether the effect was due to the fully differentiated cells or to the partially differentiated cells (or perhaps to both acting together) in the now mixed preparation.

F. Microbial Contamination: Stem cell preparations originally isolated from humans and expanded in vitro may also be variably contaminated with human viruses, bacteria, fungi, and mycoplasma. ESC preparations isolated using mouse feeder cell layers might also be contaminated with mouse viruses. Specific tests need to be performed on the source tissue and periodically on the resulting stem cell preparations to rule out the presence of these contaminants. Some of these contaminants can also multiply when stem cells are grown in vitro, and their presence can influence the results obtained when stem cell preparations are used in subsequent experiments. The presence of such contaminants can also potentially affect the reproducibility of the results of experiments in which stem cell preparations are studied in vivo in experimental animals.

In summary, there are numerous challenges to obtaining and preserving the uniform and stable preparations of stem cells necessary for reliable research and, eventually, for safe and effective possible therapies. Researchers must address multiple factors in order to maximize the probability of obtaining reproducible results with human stem cell preparations. Human stem cell preparations that are "single cell cloned," with a normal chromosome structure and number, and stored as multiple samples that are preserved at very low temperature, and compared in experiments where cells from the same lot of frozen material are used, and well-characterized as to the absence of cellular, viral, bacterial, fungal, and mycoplasma contaminants, and tested to determine the proportion of stem cells *and* various differentiated cells in the cell preparation used in the experiments, are most likely to yield experimental results that will be reproducible. Preparations with these properties will be the most useful both in basic research and in investigations of possible clinical applications.

Major Examples of Human Stel Cells

In this section we discuss major examples of human

stem cells that meet many of the criteria listed above. Among human adult stem cells, we focus on mesenchymal stem cells *(MSCs)*, multipotent adult progenitor cells (MAPCs), and neural stem cells, and among human embryonic stem cells, on ESC and EGC cells. For infor*mation on* the wide variety of other human stem cell preparations isolated from adult tissues.

Further research on some of these other adult stem cell preparations may demonstrate that they can also be "single cell cloned," expanded considerably by growth in vitro with retention of normal chromosome structure and number, and preserved by freezing and storage at low temperatures. At that point, it would be very important to compare the properties of these other adult stem cells, and the more differentiated cells that can be derived from them, with the already characterized human embryonic and adult stem cell preparations.

Human Adult Stem Cells

Human Mesenchymal Stem Cells

Bone marrow contains *at least two major kinds of stem* cells: hematopoietic stem cells, which give rise to the red cells and white cells of the blood, and mesenchymal stem cells, which can be reproducibly isolated and expanded in vitro and that can differentiate in vitro into cells with properties of cartilage, bone, adipose (fat), and muscle cells. The characteristics (morphology, expressed proteins, and biological properties) of these cells have been somewhat difficult to specify, because they appear to vary depending upon the in vitro culture conditions and the specific cell preparation. However, there is a recent report indicating that MSCs, if isolated using three somewhat different methods, give rise to stem cell preparations whose properties are very similar to one another. Using dual antibody staining and fluorescence-activated cell sorting, Gronthos and colleagues isolated human MSCs in almost pure form and expanded them substantially in vitro. Thus, human MSC preparations isolated in different laboratories by different methods may have similar but not identical properties.

A molecular analysis of genes expressed in a single-cellderived colony of MSCs provided evidence for the activity of genes also turned on in bone, cartilage, adipose, muscle, hematopoiesis-supporting stromal, endothelial, and neuronal cells. These results are surprising in that MSCs derived from a single cell appear to be expressing genes associated with multiple major cell lineages. It is possible that different cells within the colony had already entered into distinct differentiation pathways, resulting in a developmentally heterogeneous population composed of *several differen*t cell types.

Mesenchymal stem cells are important for research and therapy for several reasons. First, because they can be differentiated in vitro into multiple cell types, they make possible detailed research on the molecular events underlying differentiation into bone, cartilage, and fat cell lineages.

Second, they have recently been shown to support the in vitro growth of human embryonic stem cells. Thus, they could replace the mo*use feeder* cells used previously, obviating the need to satisfy FDA requirements for xenotransplantation, should the ESCs or their derivatives ever be used in human clinical research or transplantation therapy. Third, clinical studies are already underway in which MSCs are cotransplanted with autologous hematopoietic stem cells into cancer patients to replace their blood cell-forming system, destroyed by radiation or high dose chemotherapy. It is believed that the MSCs will support the repopulation of the bone marrow by the injected hematopoietic stem cells.

In addition, injecting allogeneic MSCs (MSCs from a genetically different human donor) may also prove valuable in modulating the immune system to make it more accepting of foreign tissue grafts.

Finally, MSCs have the potential for cell-replacement therapies in injuries involving bone, tendon, or cartilage and possibly other diseases. They are, in fact, already being tested as experimental therapies for osteogenesis imperfecta, metachromatic leukodystrophy, and Hurler syndrome. These last two studies are of great interest, since allogeneic MSCs were used and no serious adverse immune reactions were noted.

Multipotent Adult Progenitor Cells (MAPCs).

Verfaillie and coworkers recently described the isolation of MAPCs from rat, mouse, and human bone marrow. Like MSCs, MAPCs can also be differentiated in vitro into cells with the properties of cartilage, bone, adipose, and muscle cells. In addition, there is evidence for the in vitro differentiation of human MAPCs into functional, hepatocyte-like cells, a potential that has not so far been shown for MSCs. There is increasing interest in MAPCs, both as potential precursors of multiple differentiated tissues and, ultimately, for possible autologous transplantation therapy.

The relationship between human MSCs and the human MAPCs described by Verfaillie and coworkers needs to be clarified by further research. Both kinds of cells are isolated from bone marrow aspirates as cells that adhere to plastic. Each can be differentiated in vitro into cells with cartilage, bone, and fat cell properties. They express several of the same cell antigens, but are reported to differ in a few others. MAPCs have to be maintained at specific, low cell densities when grown in vitro, otherwise they tend to differentiate into MSCs. It remains important that the isolation and properties of MAPCs be reproduced in additional laboratories.

Human Neural Stem Cells.

The nervous system is made up of three major types of cells, neurons or nerve cells proper, and two kinds of supporting or glial cells (oligodendrocyte, astrocyte). Stem cells capable of differentiating into one or more of these neural cell lineages can be isolated from brain tissue (particularly the olfactory bulb and lining of the ventricles) and grown in vitro. In the presence of purified growth-factor proteins, the population of cells can be expanded by growth in vitro as round clumps of cells called neurospheres. However, many neurospheres grown in culture are developmentally heterogeneous in that they contain more than one neural cell type, and the number of self-renewing cells is frequently low (less than five percent). Although neural stem cells are still insufficiently understood, they are already proving valuable in basic research on neural development. The ability to grow reproducible neural stem cells in vitro has facilitated identification of important neural stem cell growth factors and their cellular receptors.

For example, human neural stem cells from the developing human brain cortex, expanded in culture in the presence of leukemia inhibitory factor (LIF), allowed growth of a selfrenewing neural stem cell preparation for up to 110 population doublings. Withdrawal of LIF led to decreased expression of about 200 genes, which were specifically identified through use of "gene chips" manufactured by Affymetrix. These genes are presumably involved in promoting or preserving the stem cell's capacity for self-renewal in the undifferentiated state.

The number and specificity of the molecular changes characterized in these experiments powerfully illustrate the usefulness of neural and other stem cell preparations in basic biomedical research.

Human neural stem cells are also being injected into animals to test their effects on animal models of human neurological disease. To track the fate of the introduced human cells, they must first be modified or "marked" in ways that permit their specific detection. Marked human neural stem cells are easily tracked after they are injected into experimental animals, making it possible to determine whether they survive and migrate following injection. Studies of this type have provided evidence that human neural cells can migrate extensively in the brain after injection. In addition, such cells can be injected into animal models of human diseases such as intracerebral hemorrhage and Parkinson Disease (PD) to study their effect on the progression of the disease. Although human neural stem cells may not yet be as well characterized as MSCs or ESCs, they are being actively studied with the hope that they can be used in future treatments for devastating neurological diseases such as Alzheimer Disease and PD.

Adult Stem Cells from Other Sources.

Prentice has *summarized a large amount of recent info*rmation on preparations of stem cells isolated from amniotic fluid, peripheral blood, umbilical cord blood, umbilical cord, brain tissue, muscle, liver, pancreas, cornea, salivary gland, skin, tendon, heart, cartilage, thymus, dental pulp, and adipose tissue. Studies of many of the stem cell preparations from these sources are just getting started, and further work is needed to determine their biological properties and their relatedness to other stem cell types. In some cases, the long-term expandability in vitro of these stem cells has not been demonstrated. Yet, the demonstration that they can be isolated from such tissue compartments in animals should spur the search for similar human stem cell types.

As Prentice also reports, many attempts have already been made using various preparations of adult stem cells to influence or alter the course of diseases in animal models.

Despite the fact that the stem cell preparations used are not well characterized, and reproducible results have yet to be obtained, preliminary findings are sometimes encouraging. It is of course not yet clear whether the injected cells are functioning as stem cells, fusing with existing host cells, or stimulating the influx of the host's own stem cells into the target tissue. But, if reproduced, these preliminary findings may point the way to future therapies, even in the absence of precise knowledge of the mechanism(s) of cellular action.

Human Embryonic Stem Cells 1. Human Embryonic Stem Cells (ESCs).

Human embryonic *stem cells have been isolated from the* inner cell masses of blastocyst-stage human embryos in multiple laboratories around the world. There is great interest in understanding the properties of these cells because they hold out the promise of being able to be differentiated into a large number of different cell types for possible cell therapies, as contrasted with the more limited number of cell types available by differentiation of specific adult stem cell preparations. As of July 2003, 12 ESC preparations (up from 2 such preparations a year earlier) out of a total of 78 "eligible" preparations of human ESCs were available for shipment to recipients of U.S. federal research grants. The review by Ludwig and Thomson lists more than 40 peer-reviewed human ESC primary research papers that have been published since the initial publication in 1998.

Although isolated from different blastocyst-stage human embryos in laboratories in different parts of the world, ESCs have a number of properties in common. These include the presence of common cell surface antigens (recognized by binding of specific antibodies), expression of the enzymes alkaline phosphatase and telomerase, and production of a common gene-regulating transcription factor known as Oct-4. At least 12 different preparations of ESCs have been expanded by growth in vitro, frozen and stored at low temperature, and at least partially characterized. Some of these ESC preparations have been "single-cell cloned." Human ESCs have been differentiated in vitro into neural (neurons, astrocytes, and oligodendrocytes), cardiac (synchronously contracting cardiomyocytes), endothelial (blood vessels), hematopoietic (multiple blood cell lineages), hepatocyte (liver cell), and trophoblast (placenta) lineages. In the case of neural and cardiac lineages, similar results have been obtained in different laboratories using different preparations of ESCs, thus fulfilling the "reproducible results" criterion described above. For other lineages, the results described have not yet been reproduced in another laboratory.

Embryonic Germ Cells

Human embryonic germ cells *are isolated from the primordial* germ tissues of aborted fetuses. Gearhart has summarized the results of recent research with human and mouse EG cells. One study focused on regulation of imprinted genes in EG cells: it showed "that general dysregulation of imprinted genes will not be a barrier to their (EG cell) use in transplantation studies." In addition, Kerr and coworkers showed that cells derived from human EG cells, when introduced into the cerebrospinal fluid of rats, became extensively distributed over the length of the spinal cord and expressed markers of various nerve cell types. Rats paralyzed by virus-induced nerve-cell loss recovered partial motor function after transplantation with the human cells. The authors suggested that this could be due to the secretion of transforming growth factor-á and brain-derived growth factor by the transplanted cells and subsequent enhancement of rat neuron survival and function.

Until recently, work with human EG cells came primarily from one laboratory. Recently the isolation and properties of human EG cells have been independently confirmed. Because human EG cells share many (but not all) properties with ESCs, these cells offer another important avenue of inquiry.

Embryonic Stem Cells from Cloned Embryos (Cloned ESCs).

Although it has yet to be accomplished in practice, somatic cell nuclear transfer (SCNT) could create cloned human embryos from which embryonic stem cells could be isolated that would be genetically virtually identical to the person who donated the nucleus for SCNT: hence cloned ESCs. In theory, using such cloned embryonic stem cells from individual patients might provide a way around possible immune rejection, though in practice this could require individual cloned embryos for each prospective patient—a daunting task. And clinical uses might require a separate FDA approval for every single cloned stem cell line or its derivatives.

The ability to produce cloned mouse stem cells and genetically modify them in vitro has made possible an experiment demonstrating the potential of cloned human embryonic stem cells in the possible future treatment of human genetic diseases. Rideout et al. used a mutant *mouse* strain that was deficient in immune system function. They produced a cloned mouse embryonic stem cell line carrying the mutation, and then specifically repaired that gene mutation in vitro. The repaired cloned stem cell preparation was then differentiated in vitro into bone marrow precursor cells. When these precursor cells were injected back into the genetically mutant mice, they produced partial restoration of immune system function.

Production of cloned human embryonic stem cell preparations remains techni*cally* very difficult and ethically controversial. Recently however, Chen and coworkers have reported that fusion of human fibroblasts with enucleated rabbit oocytes in vitro leads to the development of embryolike structures from which cell preparations with properties similar to human embryonic stem cells can be isolated. This work needs to be confirmed by repetition in other laboratories.

In addition, further work is needed to decisively settle the question of whether rabbit (or human egg donor) mitochondrial DNA and rabbit (or human egg donor) mitochondrial proteins persist in the embryonic stem cell preparations. Persistence of these foreign mitochondrial proteins in these human ESClike preparations could possibly increase the probability of immune rejection of the cloned cells, thus limiting their clinical application, although the immune reaction might not be as severe as that to foreign proteins produced under the direction of chromosomal genes. The presence of foreign or aberrant mitochondria also carries the risk of transmitting mitochondrial disease (caused by defects in mitochondrial DNA) that could be detrimental to the cells and to the recipient into whom they might eventually be transplanted.

Basic Research Using Human Stem Cells

Human stem cells are proving useful in basic research in several ways. They are useful in unraveling the complex molecular pathways governing human differentiation. For example, because ESCs can be stimulated in vitro to produce more differentiated cells, this transition can be studied in greater detail and under better-controlled conditions than it can be in vivo. In the best circumstances, these differentiated cells can be grown as largely homogeneous cell populations, and their gene expression profiles can be compared in detail.

Also, stem cell preparations can be used to produce populations of specialized cells that are not easily obtained in other ways. In one case, for example, this approach has provided large quantities of human trophoblast-like cells that have not been previously available. In addition, cultures of differentiated cells derived from stem cells could be used to test new drugs and chemical compounds for toxicity and mutagenicity. As experience with these differentiated derivatives of human ESCs grows, it may become possible to reduce or eliminate the use of live animals in such testing protocols.

In the near future, the differentiated state of various human cell types will be characterized not just by a few biological markers, but by the pattern and levels of expression of hundreds or thousands of genes. Integration of this knowledge with the catalog of all human genes produced during the Human Genome Project will gradually give us knowledge of which genes are key regulators of human development and which genes are central to maintaining the stem cell state. Increased understanding of the molecular pathways of human cell differentiation should eventually lead to the ability to direct in vitro differentiation along pathways that yield cells useful in medical treatment. In addition, when the normal range of gene expression patterns is known, researchers can then determine which genes are expressed abnormally in various diseases, thus increasing our understanding of and ability to treat these diseases.

A group of stem cell researchers has recently outlined a set of important research questions that, once answered, will greatly enhance our understanding of human embryonic stem cells and their potential fates and possible uses. They include the following:

- What is the most effective way to isolate and grow ESCs?
- How is the self-renewal of ESCs regulated?
- Are all ESC lines the same?
- How can ESCs be genetically altered?
- What controls the processes of ESC differentiation?
- What new tools are needed to measure ESC differentiation in vitro and in vivo?

Human Stem Cells and the Treatment of Disease

A major goal of stem cell research is to provide healthy differentiated cells that, once transplanted, could repair or replace a patient's diseased or destroyed tissues. In pursuit of this goal, one likely approach would start by isolating stem cells that could be expanded substantially in vitro. A large number of the cultivated stem cells could then be stored in the frozen state, extensively tested for safety and efficacy as outlined above, and used as reproducible starting material from which to prepare differentiated cell preparations that will express the needed beneficial properties when they are transplanted into patients with specific diseases or deficiencies.

To make more concrete both the potential of this approach and the obstacles it faces, we will summarize, as a case study example, some current information on the properties of cells derived from human stem cell populations that have been used in an animal model of Type-1 diabetes. But before doing so, we discuss an obstacle to any successful program of stem cellbased transplantation therapy: the problem of immune rejection of the transplanted cells.

Will Stem Cell-Based Therapies Be Limited by Immune Rejection?

Much of the impetus for human stem cell research comes from the hope that stem cells (or, more likely, differentiated cells derived from them) will one day prove useful in cell transplantation therapies for a variety of human diseases. Such cell transplantation would augment the current practice of whole organ transplantation. To the extent that the healing process works with in vitro derived cells, the need for organ donors and long waiting lists for organ donation might be reduced or even eliminated.

Will the recipient (patient) accept or reject the transplanted human cells? In principle, the problem might seem avoidable altogether: adult stem cells could be obtained from each individual patient needing treatment. They could then be grown or modified to produce the desired (autologous and hence rejection-proof) transplantable cells. But the logistical difficulties in processing separate and unique materials for each patient suggest that this approach may not be practical.

The cost and time required to produce sufficient numbers of well-characterized cells suitable for therapy suggest that it will be cells derived from one or another unique stem cell line that will be used to treat many (genetically different) individual patients (allogeneic cell transplantation).

When allogeneic org*an or tiss*ue transplantation is currently done using, for example, bone marrow, kidney, or heart, powerful immunosuppressive drugs—carrying undesirable side effects—must be used to prevent immunological rejection of the transplanted tissue. Without such immunosuppression, the patient's T-lymphocytes and natural killer (NK) cells recognize surface molecules on the transplanted cells as "foreign" and attack and destroy the cells. Also, in whole organ transplantation, donor T-lymphocytes and NK cells, entering the recipient with the transplanted organ, can also destroy the tissues of the transplant recipient (called "graft versus host" disease).

Are the differentiated derivatives of human stem cells as likely to incite immune rejection, when transplanted, as are solid organs? Do their surfaces carry those protein antigens that will be recognized as "foreign"? Experiments have been done to examine human ESC and MSC preparations growing in vitro for the expression of surface molecules known to play important roles in the immune rejection process. Drukker and coworkers showed that embryonic stem cells in vitro express very low levels of the immunologically crucial major histocompatibility complex class I (MHC-I) proteins on their cell surface. The presence of MHC-I proteins increased moderately when the ESCs became differentiated, whether in vitro or in vivo. A more pronounced increase in MHC-I antigen expression was observed when the ESCs were exposed to gamma-interferon, a protein produced in the body during immune reactions. Thus, under some circumstances, human ESC-derived cells can express cell surface molecules that could lead to immune rejection upon allogeneic transplantation.

Similarly, Majumdar and colleagues showed that human mesenchymal stem cells in vitro express multiple proteins on their cell surfaces that would enable them to bind to, and interact with, T-lymphocytes. They also observed that gammainterferon increased expression of both human leukocyte antigen (HLA) class I and class II molecules on the surface of these MSCs. These results indicate that it will probably not be possible to predict, solely on the basis of in vitro experiments, the likelihood that transplanted allogeneic MSCs would trigger immune rejection processes in vivo.

Many further studies in this area are badly needed. At this time there is insufficient information to determine which, if any, of the approaches to get around the rejection problem will eventually prove successful.

4

Regulation of Biotechnology: Model Framework

COORDINATED FRAMEWORK FOR REGULATION OF BIOTECHNOLOGY

This notice describes the comprehensive federal regulatory policy for ensuring the safety of biotechnology research and products. Specifically addressed are agency policies that formed part of the previously proposed Coordinated Framework for the Regulation of Biotechnology, published in the Federal Register December 31, 1984 (49 FR 50856, hereinafter "the December 84 Notice"). These agency policies build upon experience with agricultural, pharmaceutical, and other commercial products developed by traditional genetic modification techniques.

Existing statutes provide a basic network of agency jurisdiction over both research and products; this network forms the basis of this coordinated framework and helps assure reasonable safeguards for the public. This framework is expected to evolve in accord with the experiences of the industry and the agencies, and, thus, modifications may need to be made through administrative or legislative actions.

The application of traditional genetic modification techniques is relied upon bradly for enhanced characteristics of food (e.g., hybrid corn, selective breeding), manufactured food (e.g., bread, cheese, yogurt), waste disposal (e.g., bacterial sewage treatment), medicine (e.g., vaccines, hormones), pesticides (e.g. *Bacillus thuringiensis*) and other uses. Federal agencies implement an array of laws which seek to ensure the safety of these products. A concise index of these U.S. laws was published in the Federal Register November 14, 1985 (50 FR 47174, hereinafter "the November 85 Notice"). These laws are product-specific because they regulate certain product uses, such as foods or pesticides. This approach provides the opportunity for similar products to be treated similarly by particular regulatory agencies.

Biotchnology also includes recently developed and newly emerging genetic manipulation technologies, such as recombinant DNA (rDNA), recombinant RNA (rRNA) and cell fusion, that are sometimes referred to as genetic engineering. While the recently developed methods are an extension of traditional manipulations that can produce similar or identical products, they enable more precise genetic modifications, and therefore hold the promise for exciting innovation and new areas of commercial opportunity.

Concerns were raised as to whether products resulting from the recently developed techniques would pose greater risks than those achieved through traditional manipulation techniques. For example, what might be the possible environmental consequences of the many anticipated agricultural and environmental applications that will take place outside the physical contraints of a contained facility? In particular, the environmental application of genetically engineered microorganisms may elicit concern because they are of microscopic size, and some may be able to reproduce, proliferate, and become established.

The underlying policy question was whether the

regulatory framework that pertained to products developed by traditional genetic manipulation techniques was adequate for products obtained with the new techniques. A similar question arose regarding the sufficiency of the review process for research conducted for agricultural and environmental applications.

The Administration, recognizing its responsibility to confront these concerns, formed an interagency working group under the former White House Cabinet Council on Natural Resources and the Environment in the spring of 1984. The working group sought to achieve a balance between regulation adequate to ensure health and environmental safety while maintaining sufficient regulatory flexibility to avoid impeding the growth of an infant industry.

Upon examination of the existing laws available for the regulation of products developed by traditional genetic manipulation techniques, the working group concluded that, for the most part, these laws as currently implemented would address regulatory needs adequately. For certain microbial products, however, additional regulatory requirements, available under existing statutory authority, needed to be established.

The existing health and safety laws had the advantage that they could provide more immediate regulatory protection and certainty for the industry than possible with the implementation of new legislation. Moreover, there did not appear to be an alternative, unitary, statutory approach since the very broad spectrum of products obtained with genetic engineering cut across many product uses regulated by different agencies.

Because of the rapid growth in the scientific knowledge base, the working group felt strongly that the federal agencies needed to have an interagency mechanism for sharing scientific information related to biotechnology, particularly information on research and product applications submitted to the agencies.

The December 1984 Notice described the regulatory framework envisioned by the working group, and recognizing the evoluntionary nature of its development, asked for comments. In summary, the Notice stated that the Food and Drug Administration (FDA) would regulate genetic engineering products no differently that those achieved through traditional techniques. The Environmental Protection Agency (EPA) described existing and proposed new policies for regulating pesticidal and nonpesticidal microorganisms. The Department of Agriculture (USDA) stated that under its different legislative authorities it could broadly regulate genetically engineered plants and animals, and plant and animal pathogens. The Notice also proposed an interagency science coordinating mechanism.

Many comments were received in response to the Notice. These contributed to the refinement of both the regulatory requirements and the interagency science coordintion mechanism.

The interagency coordination merchanism, the Biotechnology Science Coordinating Committee (BSCC), discussed in more detail in section C, of this Preamble, came into being while the agencies were still in process of refining their regulatory proposals. Consequently, the BSCC was able to play a helpful role in the formulation of two basic principles: (1) Agencies should seek to adopt consistent definitions of those genetically engineered organisms subject to review to the extent permitted by their respective statutory authorities; and, (2) agencies should utilize scientific reviews of comparable rigor.

The regulatory framework anticipates that future

scientific developments will lead to further refinements. Experience with earlier basic scientific research has shown that as the science progressed and became better understood by the public, regulatory regimens could be modified to reflect more complete understanding of the potential risks involved. Similar evolution is anticipated in the regulation of commercial products as scientists and regulators learn to predict more precisely particular product use that require greater or lesser controls or even exemption from any federal review.

This framework has sought to distinguish between those organisms that require a certain level of federal review and those that do not. This follows a traditional approach to regulation. Within agriculture, for example, introductions of new plants, animals and microorganisms have long occurred routinely with only some of those that are not native or are pathogenic requiring regulatory approval. It should be noted that microorganisms play many essential and varied roles in agriculture and the environment and that for decades agricultrual scientists have endeavored to exploit their advantages through routine exerimentation and introduction into the environment; and as a rule these agricultural and environmental introductions have taken place without harm to the environment.

ENVIRONMENTAL PROTECTION AGENCY (EPA)

Statement of Policy; Microbial Products Subject to the Federal Insecticide, Fungicide, and Rodenticide Act and the Toxic Substances Control Act

This notice describes how EPA is addressing certain microbial products of biotechnology under the Federal Insecticide, Fungicide, and Rodenticide Act (FIFRA) and the Toxic Substances Control Act (TSCA). The notice outlines

EPA's plan for review of microbial pesticides under FIFRA with particular emphasis on small-scale field testing of genetically engineered, nonindigenous, and pathogenic microbial pesticides. It also announces EPA's policy for addressing new microbial products that fall under TSCA authority. This includes EPA's interpretation of the new chemical premanufacture notification (PMN) provisions of TSCA section 5 for new genetically engineered microorganisms used for commercial purposes, and the Agency's intentions to develop, under TSCA, a significant new use rule for pathogenic microorganisms; a rule modifying the PMN research and development exemption so that small scale field testing of microorganisms for TSCA purposes is subject to PMN; a section 8(a) reporting rule for other microorganisms prior to their release in the environment; and section 5(h)(4) exemptions as appropriate.

Information submitted as comments on this EPA notice may be claimed confidential by marking any part or all of that information as "confidential Business Information." Information so marked will not be disclosed except in accordance with procedures set forth in 40 CFR Part 2. A sanitized copy of any material containing Confidential Business Information must be provided by the submitter for inclusion in the public record. Information not marked confidential may be disclosed publicly by EPA without prior notice.

I. Overview

A. Purpose

For centuries, humans have used organisms to generate commercial products or to perform useful functions. During the last decade, advances in the biological sciences have increased the ability of humans to change or combine the inherited characteristics of microorganisms, plants, and animals. These advances, along with more traditional genetic engineering and biological techniques, are expected to lead to a wide variety of useful products. Among these are microorganisms that will be used to degrade toxic pollutants, leach minerals, enhance oil recovery, produce industrial chemicals, and act as pesticides. As with chemicals used for the smae types of purposes, many of these microorganisms will be reviewed by EPA for potential health and environmental risks.

Specifically, EPA reviews and may register pesticide products under the Federal Insecticide, Fungicide, and Rodenticide Act (FIFRA), and reviews chemical substances (except those used as pesticides, foods, food additives, cosmetics, drugs, and medical devices) under the Toxic Substances Control Act (TSCA). EPA's Office of Pesticides and Toxic Substances (OPTS) is responsible for implementing both FIFRA and TSCA.

This notice describes how EPA plans to address microbial products that are subject to FIFRA and TSCA, and explains the scope of coverage and procedures for review of these products under both statutes. The following questions are addressed in this notice:

1. What microbial products are subject to review under FIFRA and how will they be reviewed? (Unit II)
2. What microbial products are subject to review under TSCA and how will they be reviewed? (Unit III)
3. What definitions will be used to identify the products that will be addressed by the appropriate statute? (Unit IV)

In reviewing products, the Agency is required under both FIFRA and TSCA to consider the potential benefits to society as well as any potential risks. EPA will take both risks and benefits into account in its regulatory decisions concerning these products, and will implement the two statutes in as consistent a fashion as possible within statutory constraints.

B. Background

1. *December 1984 proposal.* EPA issued for comment a "Proposed Policy Regarding Certain Microbial Products" as part of the Office of Science and Technology Policy's "Proposal for a Coordinated Framework for Regulation of Biotechnology." This proposal was published in the Federal Register of December 31, 1984 (49 FR 50880) and is hereafter referred to as the "December 84 notice." Briefly, in the December 84 notice EPA proposed a mechanism for review of genetically engineered and nonindigenous microbial pesticides under FIFRA. It also described how EPA proposed to address certain genetically engineered microorganisms subject to the new chemical substance premanufacture notification (PMN) provisions of section 5 of TSCA.

2. *Comments on the December 84 notice.* EPA received comments on the December 84 notice from 68 organizations and individuals. All the comments received by EPA are available for review and copying from 8 a.m. to 5 p.m. Monday through Friday, except legal holidays, in the TSCA Public Information Office, Rm. E-107, Environmental Protection Agency, 401 M St., SW., Washington, DC 20460.

The Agency has carefully evaluated these comments. Several of the proposed policies set forth in the December 84 notice have been revised or clarified in this notice in response to these comments and as a result of the regulatory experience EPA has gained over the past year.

One of the most frequent comments addressed EPA's authority under TSCA and FIFRA. The Agency has continued to evaluate the extent and limit of its statutory authority and has concluded that TSCA and FIFRA provide sufficient authority for the Agency to meet its goals and responsibilities in regulating biotechnology products. However, some new regulations will be required and others will have to be modified in order to fully implement certain aspects of EPA's policies. These regulations and modifications are discussed in Units II and III of this notice.

Numerous commenters addressed the scope of EPA's policy and raised questions about which microbial products are subject to TSCA and FIFRA. In Units II.B, and III.B, the Agency provides detailed explanations of which microorganisms are and are not subject to FIFRA and TSCA, and from among the products that are subject, which are subject to regulatory review prior to any environmental application.

Many commenters expressed concern that the Agency was relating a microorganism's potential for risk to the process by which it was made, particularly in the definition of which microorganisms are "new" and therefore subject to PMN under TSCA. First, commenters suggested that the process by which an organism was modified was too indirect as an indicator of its newness.

They pointed out that while certain processes can be used to produce new combinations of traits in microorganisms, their use does not necessarily mean that new combinations of traits have been formed. Second, the process-based approach was believed to be an insufficient indicator of risk, because genetic engineering processes do not necessarily produce organisms that present risks, nor are non-engineered organisms necessarily safe. Finally, because the process-based approach would single out certain techniques for regulation, it would result in market distortions that favored the more traditional techniques even though the newer techniques could be as safe or safer.

After reviewing the comments, the Agency considered a number of alternatives to the "process-based" approach. In choosing among these alternatives, EPA carefully considered how well the options approximated risk (there was uncertainty with all the options in this respect), whether they could be implemented and enforced through criteria that were unambiguous to all affected persons, and (in the case of organisms subject to TSCA) the TSCA mandate to review "new" substances. The alternative EPA has chosen gives particular attention, under both FIFRA and TSCA, to microorganisms that (1) are used in the environment, (2) are pathogenic or contain genetic material from pathogens, or (3) contain new combinations of traits (e.g., organisms that are genetically modified to contain genetic material from dissimilar source organisms and organisms that are nonindigenous). EPA believes these categories have sufficiently high potential for widespread exposure, adverse effects, or uncertainty concerning potential effects to deserve particular regulatory scrutiny. This approach takes a significant step towards separating products on the basis of potential risk.

The Agency also received comments on the information and data to be submitted by companies filing notifications of intent to conduct field tests with certain microbial pesticides. These requirements have been clarified and additional references have been cited in the FIFRA unit of this notice that should provide useful guidance on what information to submit. The TSCA unit contains similar guidance on the submission of information.

Finally, several commenters addressed issues pertaining to confidential business information (CBI). Some expressed concern that CBI be adequately protected from disclosure, while others stressed the need for public access to information on new biotechnology products. EPA has summarized its position with respect to CBI and public disclosure later in this overview (Unit I.G).

A background document providing more detail on the Agency's response to comments on the December 84 notice has been placed in the public record for this notice and is available in the TSCA Public Information Office (address listed in Unit VI of this notice).

C. Summary of EPA Policy

This notice focuses on oversight and review procedures for microorganisms that are subject to FIFRA or TSCA. Microorganisms intended for use as pesticides are subject to FIFRA, and many microorganisms intended for general commercial and environmental applications (e.g., metal leaching, pollutant degradation, enhanced nitrogen fixation) are subject to TSCA. This notice addresses the rationale for various requirements and provides guidelines for compliance.

Specifically, EPA's policies that apply to microbial products subject to FIFRA or TSCA jurisdiction will include the following specific requirements:

1. Microorganisms deliberately formed to contain genetic material from dissimilar source organisms (inter-generic) will be subject to review before any environmental releases, including small-scale field testing and other environmental research and development (R&D). Under the statute, those that are subject to TSCA and used in closed systems (i.e., never intentionally released to the environment) must be reported before they are manufactured for

non-R&D commercial purposes. However, EPA is considering promulgating a rule to exempt certain contained uses from this requirement.

2. Microorganisms formed by genetic engineering other than inter-generic combinations will be subject to the following provisions: (a) if any source organism is a pathogen, the resulting microbial products are subject to review under FIFRA or TSCA prior to any environmental release, except if used solely for non-pesticidal agricultural uses, in which case they are subject only to U.S. Department of Agriculture (USDA) review (see the USDA notice in this Federal Register) (b) if source organisms are not pathogens, the resulting microbial products are subject to abbreviated review under FIFRA (if they are pesticides) before any small-scale environmental release, or will be subject to the reporting requirements of sections 8 (a) and (e) of TSCA.
3. Nonengineered microorganisms: (a) indigenous pathogens will be reviewed under FIFRA or TSCA prior to use on greater than 10 acres of land and greater than 1 acre of water, except those that are solely for non-pesticidal agricultural purposes, which will be subject only to USDA authority; (b) nonindigenous pathogens will be reviewed under FIFRA prior to any environmental release, and under TSCA prior to release at greater than 10 acres, unless they are pathogens used solely for non-pesticidal agricutural purposes in which case they will be reviewed by USDA (see USDA notice in this Federal Register); (c) nonindigenous microbial pesticides that are not pathogens will be subject to abbreviated review under FIFRA before any small scale environmental release; (d) indigenous microbial

pesticides that are not pathogens will be reviewed under FIFRA prior to use on greater than 10 acres.

4. All other microorganisms used or intended for use as pesticides and not covered in Unit I.C. 1 through 3, regardless of source, mode of action, or method of manufacture will be reviewed under FIFRA prior to use on greater than 10 acres unless exempted by regulation.
5. Manufacturers and importers of microorganisms under TSCA, if they are not otherwise subject to review, will be required to submit general information, before environmental release, that the Agency can use to monitor environmental uses and to determine if additional requirements are necessary in the future. EPA will gather such information by means of a TSCA section 8(a) reporting rule.
6. Manufacturers and importers of all microorganisms subject to TSCA must report any information on substantial risks under TSCA section 8(e). Registrants of microbial pesticides must report any information regarding unreasonable adverse effects of the pesticide on the environment under FIFRA section 6(a)(2).

A table at the end of Unit I summarizes the policies for prior notification and review of microorganisms applied in the environment.

This policy is immediately effective for microbial pesticides under FIFRA and for "new" microorganisms subject to premanufacture notification under TSCA. Implementing other aspects of the policy for TSCA substances, however, will require rulemaking. Until final rules are effective, EPA expects manufacturers to comply with most aspects of the policy voluntarily. The one exception is that manufacturers of microorganisms, described in Unit I.C.5, that are excluded from other TSCA notification requirements are not expected to report until a final section 8(a) rule is promulgated.

This notice also describes the types of information EPA expects to receive from persons subject to these policies to permit an evaluation of possible risks. EPA will determine specific information needs on a case-by-case basis, and will frequently use non-Agency experts with specific knowledge of the relevant microorganisms and uses to assist in reviews. In addition, EPA is establishing a biotechnology Science Advisory Committee (SAC) to provide peer review of specific cases and advice on technical issues. The SAC will be composed of non-Agency scientists and members of the lay public. More information on the SAC may be found in Unit I.F.

Although many of the policies described in this notice are immediately effective, the Agency recognizes that biotechnology is a repidly developing field and that newly available information may affect the judgments underlying these policies. Accordingly, EPA recognizes that modifications of these policies may be necessary in the future, and it is willing to make such modifications as may be appropriate. Therefore, EPA encourages all interested persons to provide comments on the policies described in this notice. Comments should be submitted to the address provided at the beginning of this EPA notice. The public will have additional opportunities for comment when the Agency proposes rules for those parts of its policy that require rulemaking procedures. Thse parts are specifically indicated in Units II and III.

D. Rationale for Approach

This unit provides a discussion of EPA's rationale for

giving special focus to environmental release, pathogens, and microorganisms with new characteristics (e.g., containing genetic material from dissimilar source organisms or nonindigenous organisms).

1. Environmental releases. Physical containment can be used to mitigate undesirable or unexpected characteristics of a microorganism by providing the means to control a microorganism's growth, reproduction, and exposure to other organisms. However, microorganisms meant to be released in the environment are not subject to this control mechanism. Although many microorganisms will be biologically contained, that is, they will have existing and inherent limitations on their growth and survival, some of them may reproduce and thereby increase in number in the environment beyond the amounts originally released. Some will also have independent mobility, or may be spread beyond the area in which they are used. Thus, to ensure that environmental releases of microorganisms do not pose unreasonable adverse effects, the Agency has determined that it should review and evaluate proposals for certain environmental releases before they are allowed to proceed. The microorganisms to be subject to review before any environmental release are described in the following paragraphs, and in Units II and III of this notice.

The Agency acknowledges the difficulty of defining environmental release. For now, the Agency's approach will focus on when an organism is considered to be contained rather than when it is released. Guidance is provided in Unit IV on how to determine whether a microorganism is considered to be contained. The definition of environmental release will be refined in subsequent rulemaking activities.

2. *Pathogenic microorganisms.* Given their ability to cause disease in plants, animals, humans, and microbes,

EPA generally believes pathogenic microorganisms should be reviewed before they are released in the environment.

As used in this notice, a "pathogen" is a microorganism that has the ability to cause disease in living organisms. This includes previously documented pathogens, and microorganisms deliberately formed to contain genetic material from pathogens (e.g., through genetic engineering techniques). A complete discussion of the definition of pathogenicity is included in Unit IV, as well as guidance to aid in the determination of whether a particular microorganism falls within the scope of the EPA policies that address pathogens.

Pathogens are a clearly defined category of organisms known to cause adverse effects. In addition, because of the increased uncertainty about behavioral changes that may be associated with genetically engineered pathogens, the Agency has decided to review genetically engineered pathogens prior to any environmental release (including small-scale field testing). However, the Agency will defer review of nonengineered indigenous pathogens until they are used in larger scale applications (greater than 10 acres), because ample experience indicates that nonengineered, indigenous pathogens are sufficiently well controlled by natural mechanisms in small-scale environmental applications. Further, the Agency will not review pathogens used solely for non-pesticidal agricultural purposes (except those formed through inter-generic combinations, which are "new") because these are adequately reviewed by the USDA (see the USDA notice in this Federal Register).

The Agency's decision to focus on pathogens does not mean that EPA has concluded that nonpathogens are necessarily safe or that all pathogens present unreasonable risks. In fact, the Agency expects to identify widely varying degrees of risk among different uses of pathogens. It should be clear that other considerations besides pathogenicity will affect the evaluation of risk, e.g., functions of the recombined genes, possibilities for genetic transfer, environmental fate, and potential competition with other organisms. When other considerations indicate that it is appropriate, the Agency will consider excluding specific categories of pathogens from review, or may provide guidance that would limit the information requirements associated with its reviews of pathogens. As explained in Unit IV, the Agency has already exempted from review as pathogens organisms that incorporate only certain genetic material from pathogens.

3. *Microorganisms with new characteristics.* A third factor that makes potential adverse effects of microorganisms less predictable is the existence of new traits or characteristics. These traits may be new to the organism, or new to the environment in which the organism is released.

a. *Microorganisms having significant potential to exhibit new traits.* Modern genetic engineering techniques permit genetic material to be intentionally combined in organisms that would not normally share that genetic material. Some of these genetically engineered microorganisms may exhibit new or altered traits affecting, for example, their survivability, host range, substrate utilization, competition with other organisms, or protein or polysaccharide production. In some cases such microorganisms may be able to evade or overcome natural controls on their growth, or controls on their ability to cause adverse effects. In many other cases, their natural hardiness will be reduced.

In addition to the possibility that certain engineered organisms may exhibit new traits, if they are released they may be transported through natural dispersal mechanisms to other areas in the environment that have not previously contained organisms having these new combinations of traits.

Because of these considerations, EPA's policies will give particular regulatory attention to organisms that have a significant probability of exhibiting a new trait or combination of traits (standards for this are explained below). This approach accomplishes two important objectives. First, it identifies a group of microorganisms whose behavior in the environment poses significant uncertainty and thus warrants regulatory review. Simultaneously, it provides a way of defining "new" microorganisms that are subject to PMN requirements under TSCA (see Unit III.C.1).

EPA's policy, specifically, focuses on microorganisms that have been deliberately altered to contain genetic material from dissimilar source organisms, because such organisms are more likely to exhibit new combinations of traits and their behavior is therefore less predictable. Given this conceptual basis, the question then becomes how dissimilar two organisms must be before combinations of genetic material between them are likely to produce "new combinations of traits."

Based on the following considerations, EPA has decided that inter-generic combinations (combinations from source organisms of different genera) but not intra-generic combinations (source organisms from the same genus) are sufficiently likely to result in new combinations of traits that they should be given special attention. First, combinations of genetic material from microorganisms from different genera are more likely to result in new traits than combinations of genes from microorganisms within the same genus. Also, while genetic exchange occurs naturally and somewhat commonly among many microorganisms, it is more likely to occur in nature within a single genus than across many different genera (Refs. 2, 12, 13). Finally, genus designations provide a practical criterion for administrative and regulatory purposes.

The Agency has decided to exclude certain combinations from special consideration as inter-generic organisms. Excluded are inter-generic combinations in which the genetic material added to the recipient microorganism consists only of well-characterized, non-coding regulatory regions. The resulting organisms do not possess new combinations of traits; rather, they exhibit quantitative changes in preexisting traits. In addition, if experience or data indicate that certain other inter-generic combinations warrant exclusion, the Agency will use the appropriate statutory or policy mechanisms under FIFRA and TSCA to waive certain requirements for reviewing them. For example, EPA is considering exempting from PMN review under TSCA those inter-generic combinations used only in physically contained systems.

Although EPA considers intra-generic combinations to be less likely to produce new combinations of traits than inter-generic combinations, the Agency realizes that science provides no absolute standard for such distinctions. Nevertheless, EPA believes the approach it has adopted is practical and facilitates the identification of those microorganisms that should be subject to special attention and also that should be considered "new" under TSCA. If experience reveals that intra-generic combinations that could cause adverse effects will be developed, the Agency will modify its policies to require review of these products.

Unit IV contains more detailed guidance for determining if a given microorganism is the result of an inter-generic combination. The determinations are based on taxonomic designations of organisms. The Agency is aware that microbial taxonomy is a dynamic and often controversial science (Refs. 4, 18) and that new information concerning microorganisms' properties and interrelationships will alter taxonomic designations. However, the Agency believes that its procedures can be sufficiently flexible to accommodate the developments that will occur, and that there are many significant advantages to using taxonomic standards. These advantages are discussed in more detail in Unit IV.

b. *Nonindigenous microorganisms.* Another category of organisms that are likely to exhibit traits new to an environment is nonindigenous microrganisms. Applications of nonindigenous microorganisms in the environment could pose a high degree of uncertainty with respect to their behavior. Experience shows that scientists cannot always accurately predict how such organisms will behave in their new environment (Ref. 15, 16). It can be difficult to predict whether a nonindigenous microorganism will be subject to the physical and biological control factors present in the environment where it is to be introduced. In a small number of cases, nonindigenous pathogens such as the chestnut blight fungus and the Dutch elm disease fungus have caused significant adverse effects. As a result, there exist today regulations that govern the intentional movement of some, but not all, nonindigenous species (e.g., the Plant Pest Act administered by USDA). EPA believes that nonindigenous microorganisms whose uses are covered by FIFRA should be subject to Agency review and evaluation before they are released in the environment, to minimize the uncertainties with respect to their behavior. However, EPA does recognize that small-scale use of certain nonindigenous microbial pesticides (i.e., pathogens) may pose greater potential risk than others, and has accordingly adopted abbreviated review procedures for small-scale use of nonpathogenic nonindigenous microbial pesticides. Unit II addresses these issues, and Unit IV provides guidance on determining whether a microorganism is nonindigenous.

E. Explanation of Jurisdiction — EPA and USDA

Both EPA and USDA seek to assure the safety of microbial products and yet minimize impediments to intellectual and economic advances in biotechnology. Because some of the statutes the agencies administer entail overlapping responsibilities, the two agencies are eliminating duplicative requirements wherever possible and coordinating their reviews.

Where allowed by statute, EPA and USDA have sought to eliminate overlapping reviews altogether. This notice reflects many instances where this has been done. Where overlaps could not be avoided, the agencies have established mechanisms for coordinating their reviews. EPA and USDA will identify principal liaisons who will have the responsibility to share information, coordinate data requests, and keep one another informed of communications with submitters. Also, the agencies will form a coordinating committee to meet periodically and work out general coordination problems that may transcend specific reviews. Finally, the National Biological Impact Assessment Program that has been established within USDA will provide a common resource of scientists available to both agencies to review procedures, protocols, and projects on an advisory basis.

Submitters are encouraged to contact either agency if they have jurisdictional questions, but general guidelines are described below.

First, inter-generic microorganisms containing genetic material from a pathogenic source organism must be reported to both agencies (definitions of "inter-generic" and "pathogen" may be found in Unit IV). In this case, statutory constraints make it necessary for both EPA and USDA to review the products because the microbes are potential "pests" subject to the Plant Pest Act, and they are "new" and therefore subject to TSCA premanufacture notification (or they are pesticides and subject to FIFRA notification). However, the agency reviews have somewhat different purposes, in that the EPA review is for a general use of an organism under TSCA or for use as a pesticide under FIFRA, while the USDA review is for a specific permit application. The agencies will coordinate these reviews as explained earlier.

Second, persons developing inter-generic organisms that contain no genetic material from a pathogen and that do not meet the USDA definition of a "plant pest" will be expected to report only to EPA; they will not report to USDA at all. EPA will inform USDA and the submitter if any data suggest that the organism has pest qualities which may require a permit from USDA. This avoids unnecessary duplication of effort and is consistent with the non-discretionary responsibility under TSCA to review new organisms and under FIFRA to review pesticides.

Third, in the case of intra-generic engineered organisms that contain genetic material from a pathogen, the use of the organism will determine which agency reviews it. When used solely for non-pesticidal agricultural purposes, such organisms must be reported only to USDA under the Plant Pest Act. When used for non-agricultural purposes, such organisms should be reported to EPA, either voluntarily under the TSCA section 5(a)(2) rule EPA will be developing or, if the organism is a pesticide, under FIFRA. In both cases, the microorganisms should also be reported to USDA as potential plant or animal pathogens. When such dual reporting is necessary, the agencies will assist the submitter by coordinating through the mechanisms described above.

In the case of intra-generic microbes containing no genetic material from pathogens and nonengineered microorganisms, EPA will gather general information under section 8(a) of TSCA and conduct abbreviated reviews under FIFRA (see Units II and III of the EPA notice). Both agencies agree that members of this category of microbes, in general, present the lowest risk and therefore do not need a high level of scrutiny before any release into the environment. However, the FIFRA abbreviated reviews and the TSCA section 8(a) reporting will ensure that both agencies are aware of environmental releases of these organisms and can take appropriate action when necessary.

F. EPA Biotechnology Science Advisory Committee

EPA is establishing a Science Advisory Committee for biotechnology. The formation of this committee is consistent with intentions stated in two Federal Register notices issued by the Office of Science and Technology Policy (49 FR 50904, December 31, 1984 and 50 FR 47174, November 14, 1985). The committee's primary functions will be to provide peer review of specific product submissions under TSCA, FIFRA, and other EPA statutes and scientific oversight of the Agency's biotechnology programs.

The committee will consist of independent scientists and members of the lay public. It will be of sufficient size and diversity to provide the range of expertise required to assess the scientific and technical issues pertinent to its responsibilities. The committee will be supplemented by consultants when they are needed to extend the range of expertise of the standing committee, and will be authorized to form subcommittees or panels for any purpose consistent with its charter.

Scientific members of the committee will be selected on the basis of their professional qualifications to examine the questions of hazard, exposure, and risk to humans, other non-target organisms, and ecosystems. Some committee members will serve as liaisons (holding joint membership) with the FIFRA Scientific Advisory Panel (SAP) and with the EPA Science Advisory Board (SAB). The SAC will also include nonvoting representatives from other Federal agencies that are involved in regulating products of biotechnology.

The Agency intends for meetings of the SAC to be open to the public. Meetings may be closed by the Chairperson when necessary, such as during discussion of issues subject to statutory confidentiality requirements, but EPA will encourage open public discussion of issues to the greatest extent possible (see unit I.G).

G. Confidential Business Information

Both FIFRA and TSCA generally prohibit the Agency from releasing certain confidential business information (CBI). These prohibitions apply to information on products of biotechnology, and the Agency will meet its obligations to protect information claimed confidential by applicants and other data submitters. However, the Agency also recognizes that there is strong public interest in many aspects of biotechnology, particularly in the possibility of adverse effects resulting from the environmental release of genetically engineered organisms. Accordingly, if is the Agency's policy to carry out as much of its review as possible in the open, in order to provide an opportunity for public participation and to increase public confidence in the review process. The Agency is encouraged by the extent to which industry and other submitters have been willing to authorize the release of relevant information to date and urges future data submitters to limit confidentiality claims as much as possible in order to foster an open review process.

H. International Aspects

EPA is committed to the policy described in the section entitled "International Aspects" in the Office of Science and Technology Policy Preamble, published in this Federal Register.

II. Applicability of the Federal Insecticide, Fungicide, and Rodenticide Act (FIFRA) to Microbial Products

A. Background

Biological agents, including microorganisms, may be used as pesticides, and as such they are subject to regulation under FIFRA unless specifically exempted by regulation. FIFRA establishes EPA's authority over the distribution, sale, and use of pesticide products. Before EPA can register a pesticide, it must have sufficient data to determine that the product, when used in accordance with widespread and commonly recognized practice, will not cause (or significantly increase the risk of) unreasonable adverse effects to humans or the environment. In recent years, the Agency has put in place policies, procedures, and regulations to address the human health and environmental concerns raised by the application of biological pesticides (including genetically engineered and nonindigenous microbial products) in the environment. This unit outlines EPA's regulatory mechanism for these products and updates its policy on small-scale field testing of microbial pesticides.

Regulations promulgated under FIFRA and appearing at 40 CFR 162.5(c)(4) specify that microorganisms, when used as pesticides, are regulated under FIFRA. The specific kinds of data and information that are required to support the registration of each microbial pesticide under FIFRA are detailed in 40 CFR 158.65, 158.170, and 162.163. The Agency has also published guidance for developing these data in the Pesticide Assessment Guidelines: Subdivision

M — Biorational Pesticides (Ref. 20).

The Agency must conduct a complete evaluation and review of the data submitted to support any pesticide registration before determining whether the pesticide should be registered. This evaluation is conducted with respect to the general criteria set forth in 40 CFR 162.7(d) and (e) and 162.167. Prior to registration, producers may test their pesticide products under an experimental use permit (EUP), issued pursuant to section 5 of FIFRA and 40 CFR Part 172. The data and information needed to support the issuance of an EUP for microbial pesticides are specified at 40 CFR 158.170.

The regulations governing EUPs include a generally applicable presumption that EUPs will not be required for certain small-scale experimental uses of new pesticides (or new uses of previously registered pesticides). Recently, however, the Agency issued a statement of interim policy on small-scale field testing of nonindigenous and genetically altered microbial pesticides, published in the Federal Register of October 17, 1984 (49 FR 40659); see also 49 FR 50882, December 31, 1984. Briefly, the policy statement announced that the small-scale field test provision of 40 CFR 172.3 would not automatically apply to, and that the Agency should be notified before the initiation of, any field testing of genetically altered or nonindigenous microbial pesticides to determine if EUPs are required. This policy is being revised by this notice and is discussed in detail in Unit II.D.

B. Scope of FIFRA

1. *Pesticides addressed by this notice.* All pesticides whose active ingredient(s) consist of microorganism(s) (i.e., all microbial pesticides) are addressed by this notice. Microbial pesticides may include bacteria and blue-green

algae, fungi, viruses, and protozoa used as pest control agents.

2. *Pesticides not addressed by this notice.* The Agency has determined that certain nonmicrobial organisms which fall within the definition of biological control agents are already addressed by other agencies, specifically USDA and the Department of the Interior. Examples of these biological control agents are vertebrates, insect predators, nematodes, and macroscopic parasites. Therefore, pursuant to section 25(b) of FIFRA and 40 CFR 162.5(c)(4), these nonmicrobial biological control agents have been exempted from regulation under FIFRA. However, if EPA, in cooperation with other agencies, determines that certain biological control agents exempted by § 162.5(c)(4) are not being adequately regulated, these organisms will be referred to the attention of the appropriate agency or added to the exceptions in § 162.5(c)(4) by amendment. In the latter case, those organisms would no longer be considered exempt from the provisions of FIFRA.

This unit of the notice does not address any chemical pesticide product or byproduct produced by microorganisms. Such chemicals are covered under current pesticide regulations, registration procedures, data requirements, and testing guidelines (see 40 CFR Parts 158 through 180; and Subdivisions D through O of the Pesticide Assessment Guidelines).

3. *Information-gathering policy.* In order to expand its level of knowledge and expertise, monitor the industry, and determine whether its current policy needs modification, the Agency needs as complete a data base as possible. Accordingly, those developing microbial products intended for use as pesticides that are not otherwise subject of FIFRA review are encouraged to keep the Agency apprised of their

activities. In addition, registrants of microbial pesticides are reminded that, pursuant to FIFRA section 6(a)(2), they must report any information regarding unreasonable adverse effects of the pesticide on the environment.

C. Microbial Pesticides — History and Long-Term Regulatory Strategy

1. *History*. Microbial pesticides have been in use for many years. In 1948, the Federal Government registered the first such product, *Bacillus popilliae,* to control Japanese beetle larvae in turf. However, it was not until the late 1960s and early 1970s that interest in microbial pesticides began to increase. At that time, EPA began to develop policies and procedures to specifically address microbial pesticide products. In 1983, EPA's Office of Pesticide Programs issued testing guidelines for microbial pesticides (Ref. 20). A year later, EPA issued a final regulation (40 CFR Part 158) specifying the data requirements for pesticide registration (including genetically engineered microbial pesticides). As of 1985, there were 14 microbial pesticides used in several hundred separate products registered for use in agriculture, forestry, mosquito control, and homes.

As indicated in Unit II.A above, EPA issued an interim policy on small-scale field testing of genetically altered and nonindigenous microbial pesticides in October 1984 (49 FR 40659). To date, under this policy, EPA has received and reviewed five notifications for genetically engineered microbial pesticides and two notifications for nonindigenous microbial pesticides. Three EUP applications, required in part to address unresolved issues identified in the review of these notifications, have since been received. These applications were for genetically engineered microbial pesticides.

2. *Long-term regulatory strategy*. Although EPA has an established regulatory mechanism for microbial pesticides,

the Agency envisions some further modifications in the future to specify certain policies in more detail, keep the assessment process current with existing scientific knowledge, and ensure an efficient review mechanism. Some of these anticipated modifications are discussed here.

As noted in Unit I, EPA intends to revise the EUP regulations (40 CFR Part 172) to incorporate the concepts embodied in the interim policy on small-scale field testing. Specifically, Part 172 will be revised to specify more clearly which applicants must notify EPA before conducting small-scale field tests with microbial pesticides and the content of notification.

As noted in the overview to this EPA notice (Unit I.F), EPA is forming a Science Advisory Committee. The Scientific Advisory Panel, an advisory group mandated by FIFRA, will continue to serve in its advisory capacity on specific submissions under FIFRA, until the SAC is formed.

FIFRA requires EPA to review and periodically update its guidelines, and OPP has begun this process for the Subdivision M Pesticide Assessment Guidelines. The Guidelines are currently being revised to reflect current testing methodology and advances in risk assessment capabilities resulting from OPP's recent experience in evaluating genetically engineered microbial pesticides. In addition, as the Agency gains risk assessment experience and assembles a larger body of risk assessment data, it may be appropriate to amend the Part 158 data requirements regulation to add to or modify the data requirements that apply to genetically engineered and nonindigenous microbial pesticides.

D. Regulatory Review of Microbial Pesticides

This unit describes EPA's data requirements and review

procedures for microbial pesticides. In particular, Unit II.D.1 describes the requirements and review plan for those microbial pesticides subject to review under FIFRA before they may be used in any application in the environment (i.e., small-scale field testing). Unit II.D.2 outlines the regulatory review for those microbial pesticides subject to the FIFRA requirements for an experimental use permit or registration. In most instances, microbial pesticides subject to the provisions in Unit II.D.1 will also be subject to the provisions in Unit II.D.2 when they are to be used for larger scale or commercial purposes in the environment.

1. *Small-scale field testing.* Prior to obtaining a registration for a pesticide product, applicants generally need to conduct field studies in order to gather product performance, use, and other types of data necessary to support the registration of their product. The regulations governing field studies (40 CFR Part 172) include a generally applicable presumption that EUPs will not be required for certain small-scale uses of new pesticides (or new uses of previously registered pesticides). The Agency issued a statement of interim policy addressing small-scale field testing of microbial pesticides in 1984. The interim policy announced that the Agency should be notified before initiation of any field testing of genetically altered or nonindigenous microbial pesticides. The purpose of this policy is to provide a mechanism for the Agency to evaluate these proposed small-scale field tests for possible risk to human health or the environment and determine whether EUPs are required before the tests can be initiated.

Small-scale field studies are (1) terrestrial field studies that involve 10 acres or less of land; and (2) aquatic field studies that involve 1 surface acre or less of water.

To minimize the regulatory burden on producers of

genetically engineered and nonindigenous microbial pesticides, and more closely correlate the level of Agency review with potential risk of the microorganism, the Agency has adopted a two-level review system based on its evaluation of the potential risks posed by various types of microorganisms. The two-level system will allow the Agency to receive some basic information on small-scale testing of genetically engineered and nonindigenous microorganisms that are less likely to pose significant risks to humans or the environment (Level I reporting), while reserving full notification and review procedures for microorganisms about which there is more concern (Level II notification). The review system is designed so that producers of microbial pesticides may proceed with their small-scale field tests without Agency approval, unless they are notified within a specified time that additional information or an EUP is required. In the case of level I reporting, producers need only provide a limited amount of information, and are assured of an expedited response from the Agency if it is determined that additional information is required.

The two-level system is based on the analysis set forth at Unit I.D, in which the Agency has defined groups of microorganisms that raise more concerns about their likelihood to pose risks to humans or the environment, when released into the environment, than other microorganisms. Specifically, these include microbial pesticides formed by deliberately combining genetic material from organisms of different genera and genetically engineered or nonindigenous microbial pesticides derived from pathogenic source organisms. However, other genetically engineered and nonindigenous microbial pesticides are less likely to pose significant risks to humans or the environment when applied in small-scale field test. Accordingly, the Agency has determined that this second category of microbial pesticides will be subjected to a reporting requirement and will be reviewed as described in Unit II.D.1 a through c below. The Agency will have up to 30 days to review the reported information. The kind of information needed to fulfill the reporting requirement is typically already available to an applicant as an essential part of product research and development, and is not generally expected to require generation of new data.

All microbial pesticides formed by deliberately combining genetic material from organisms of different genera, and all genetically engineered or nonindigenous microbial pesticides derived from pathogenic source organisms will be subject to the full notification requirements (Level II) as described in Unit II.D.1.e below. The Agency has determined that these organisms should continue to be subjected to the full notification and review procedures set out in the original interim policy published on October 17, 1984. The Agency will have up to 90 days to review a Level II notification.

The scope and requirements for Level I reporting and Level II notification are detailed below. The interim policy as revised by this notice does not apply to studies conducted under enclosed, contained conditions, as defined in Unit IV.

a. *Level I reporting.* Level I reporting for small-scale field testing applies to all genetically engineered or nonindigenous microbial pesticides not otherwise covered by Level II notification as detailed in II.D.1.d below. Small-scale field tests of additional groups of genetically engineered and nonindigenous microbial pesticides now covered by Level II notification may also be determined to warrant only abbreviated review in the future. The Agency will make these determinations on a case-by-case basis.

b. *Level I information.* Each report should include the following information, or, where specific information is not submitted, documentation of why it is not practicable or necessary to provide the information.

(1) Identity of the microorganism, including characteristics, and means and limits of detection.
(2) Description of the natural habitat of the microorganism or its parental strains, including information on natural predators, parasites, and competitors.
(3) Information on the host range of the parental strain(s) or nonindigenous microorganism.
(4) Information on the relative environmental competitiveness of the microorganism, if available.
(5) If the microorganism is genetically engineered, information should be provided on the methods used to genetically engineer the microorganism(s); the identity and location of the rearranged or inserted/ deleted gene segment(s) in question; a description of the new trait(s) or characteristic(s) that are expressed; information on potential for genetic transfer and exchange with other organisms, and on genetic stability of any inserted sequence.
(6) A description of the proposed testing program, including site location, crop to be treated, target pest, amount of test material to be applied, and method of application.

c. *Level I reporting process.* EPA with have up to 30 days to review the above information to make a preliminary determination of the need for an EUP. If the Agency does not notify the applicant of the need for an EUP within the 30 days, the applicant may proceed with the proposed field test. If, on preliminary assessment, the test raises sufficient concerns such that the Agency determines that additional information or monitoring is warranted, then an EUP will be required (e.g., microorganisms for which there is limited scientific information or regulatory experience, or that raise significant questions concerning genetic stability, competitiveness, or mode of action, or that warrant specific environmental monitoring during the test). In this case, the applicant has two options. First, the applicant may apply for a permit, providing the necessary data and information required to support the application. Alternatively, the applicant may provide all additional data and information required under Level II notification as outlined in Unit II.D.1.e below. If the latter option is chosen, the Agency will have an additional 60 days to review the full notification package and make a final determination as to whether an EUP is required.

d. *Level II notification.* Level II notification for small-scale field testing applies to microbial pesticides: Microbial pesticides formed by deliberately combining genetic material from organisms of different genera, genetically engineered microbial pesticides derived from source organisms that are pathogens (as defined in Unit IV), and nonindigenous pathogenic microbial pesticides (as defined in Unit IV).

e. *Level II requirements.* Notification should include adequate information to allow the Agency to evaluate the small-scale field testing program. Each notification should include the following information, or, where specific information is not submitted, documentation of why it is not practicable or necessary to provide the information.

(1) Background information on the microorganism.

(a) Identity of the microorganism, including tables of

characteristics, and means and limit of detection using the most sensitive and specific methods available.

(b) Description of the natural habitat of the microorganism or its parental strains, including information on natural predators, parasites, and competitors.

(c) Information on host range, especially infectivity and pathogenicity to nontarget organisms.

(d) Information on survival and ability of the microorganism to increase in numbers (biomass) in the environment (e.g., laboratory or containment facility test data).

(e) If the microorganism is genetically altered, the following information should be provided in addition to the information listed in (a) through (d) above:

i. Information on the methods used to genetically alter the microorganism.

ii. The identity and location of the rearranged or inserted/deleted gene segment(s) in question (host source, nature, base sequence data, or restriction enzyme map of the gene(s)).

iii. Information on the control region of the gene(s), and a description of the new trait(s) or characteristic(s) that are expressed.

iv. Information on potential for genetic transfer and exchange with other organisms, and on genetic stability of any inserted sequence.

v. Information on relative environmental competitiveness compared to the parental strains.

(2) Description of proposed field test.

(a) The purpose or objectives of the proposed testing.

(b) A detailed description of the proposed testing program, including test parameters.

(c) A designation of the pest organism(s) involved (common and scientific names).

(d) A statement of composition for the formulation to be tested, giving the name and percentage by weight of each ingredient, active and inert, production methods, contamination with extraneous microorganisms, potency and amount of any toxins present, and where applicable the number of viable microorganisms per unit weight or volume of the product (or other appropriate system for designating the quantity of active ingredient).

(e) The amount of pesticide product proposed for use and the method of application.

(f) The State(s) in which the proposed program will be conducted, and specific identification of the exact location of the test site(s) (including proximity to residences and human activities, surface water, etc.).

(g) The crops, fauna, flora, geographical description of sites, modes, dosage rates, frequency, and situation of application on or in which the pesticide is to be used.

(h) A comparison of the natural habitat of the microorganism with the proposed test site.

(i) The number of acres, number of structural sites, or number of animals/plants, by State, to be treated or included in the area of experimental use, and the procedures to be used to protect the test area from intrusion by unauthorized individuals.

(j) The proposed dates or period(s) during which the testing program is to be conducted, and the manner in which supervision of the program will be accomplished.

(k) A description of procedures for monitoring the microorganism within and adjacent to the test site during the field test.

(l) The method of disposal or sanitation of plants, animals, soils, etc., that were exposed during or after the field test.

(m) Means of evaluating potential adverse effects and methods of controlling the microorganism if detected beyond the test area.

In addition, the following references should be consulted for further guidance of the kinds of data and information that may be relevant to the evaluation of genetically engineered microorganisms: "Proposed Points to Consider for Environmental Testing of Microorganisms" developed by the National Institutes of Health Recombinant DNA Advisory Committee Working Group on Release into the Environment (Ref. 11); "Subdivision M: Biorational Pesticides" (Ref. 20); a report by the Cornell Ecosystems Research Center titled "Potential Impacts of Environmental Release of Biotechnology Products: Assessment, Regulation, and Research Needs" (Ref. 9); a National Science Foundation Report titled "The Suitability for Environmental Applications of Biotechnology" (Ref. 3); and EPA "Points to Consider in the Microorganisms" (available from TSCA Assistance Office at the address given at the beginning of this notice).

f. *Level II review process.* Once the supporting data have been submitted, EPA has up to 90 days to review each notification of intent to conduct small-scale field testing and to determine whether an EUP is required. The Agency encourages prospective applicants to meet with EPA prior to submission of their notification to discuss their field test and determine what specific data would be necessary to evalaute the product.

EPA's review process will include some or all of the elements described in the following paragraphs. As the Agency builds a baseline of risk assessment data and gains more experience in evaluating these products, certain steps may no longer be necessary. In addition, an abbreviated review process may be appropriate in some situations (e.g., review of a proposal that is similar to an already reviewed case). Such a determination will be made on a case-by-case basis.

Once a notification is received, OPP reviews each proposal and assesses potential risks associated with the proposed experiment. OPP develops a written scientific position for each proposal which identifies potential problems or significant unanswered questions and sets forth a statement of the overall likelihood of significant risk from the proposed field testing. As the review process proceeds, it may be necessary for OPP to request supplemental information.

OPP obtains comments on its assessment from a workgroup within EPA and from other Federal agencies as appropriate (e.g., USDA, National Institutes of Health, Food and Drug Administration, and National Science Foundation). Their comments are incorporated into the scientific position, as appropriate.

OPP contacts the appropriate State pesticides regulatory authorities to ensure that they are aware of the proposal and to discuss EPA's assessment. These contacts ensure that the actions of EPA and the State agencies are as consistent as possible. OPP also notifies the Animal and Plant Health Inspection Service (APHIS) of the USDA so that they can determine whether any aspect of the proposed experiment falls within APHIS jurisdiction and, if so, to avoid duplicative or conflicting assessments.

Thus far, reviews of small-scale field testing proposals for genetically engineered microbial pesticides have emphasized some questions that have not been as significant in the assessments of naturally occurring microbial pesticides. For example, OPP has identified potential risks associated with the transfer of inserted genetic material to other organisms, the competitiveness of the engineered organism compared with the parental organisms in the environment, and the ability of the engineered organism to become established in a new ecological niche and thereby pose a potential adverse environmental impact.

OPP has addressed these and similar questions on a case-by-case basis in its risk assessments. In some cases, applicants have addressed questions by redesigning the proposed application or test microorganism to minimize the potential risk. In other instances, EPA has established data requirements and test methods as a baseline, and has designed specific laboratory test(s) (or tiered series of tests) to establish whether the effect of concern is likely to materialize under field conditions.

If the notification raises complex or controversial scientific questions, OPP provides the notification package and its scientific evaluation to a group of independent scientists constituted as a subpanel of FIFRA's Scientific Advisory Panel. Separate subpanels may be formed to review each proposal since each microorganism and its proposed use may differ and raise questions that require the analysis of individuals with different expertise. The purpose of the SAP subpanel is to obtain an independent peer review of the OPP scientific position, to address specific scientific questions raised by OPP, and to identify any additional points, questions, or problems. As noted previously in Unit I.F, the Agency is forming a Science Advisory Committee which will assume these responsibilities in the future.

At the conclusion of the review, the Agency then decides whether an EUP is required. The decision document sets forth OPP's conclusions with respect to potential risks associated with the proposal, identifies any remaining questions or additional data that may be needed to complete the risk assessment, and, if an EUP is required, may recommend restrictions, limitations, or modifications of the proposal to address areas of concern. If an EUP is not required, the applicant may proceed with the proposed field test. If an EUP is required, the applicant must apply for a permit, providing the necessary data and information required to support the application. The Agency may decide to require an EUP to ensure that the experiment is conducted within certain defined limitations, the necessary data are developed to assess the proposal, or certain kinds of data are developed during the test and reported to the Agency.

2. *EUPs, large-scale testing, and registration.* Before a pesticide may be marketed as a commercial product, it must first be registered as provided for in section 3 of FIFRA. Large-scale field testing of a microbial pesticide is often necessary to evaluate a potential product and obtain data needed to support registration of the product. This testing, like small-scale field testing under an EUP, is subject to section 5 of FIFRA which authorizes EPA to approve applications for EUPs for limited use of an unregistered product or use of a registered product for an unregistered use. Data requirements for registration are specified in 40 CFR 158.170 and a subset of these requirements applies to large-scale field testing proposals to be performed under EUPs. The regulatory review process

consists of the same basic elements in both situations and is described in this unit.

a. *Scope.* All microbial pesticides to be used in large-scale field tests are subject to review under FIFRA EUP regulations. The conditions under which an EUP is required are specified at 40 CFR Part 172, which also provides guidance on how to determine whether an EUP must be obtained. Likewise, all microbial pesticides are subject to the FIFRA registration requirements.

b. *General requirements for microbial pesticides.* The existing pesticide data requirements and regulations governing large-scale field testing (40 CFR Parts 158 and 172) and registration (40 CFR Parts 158 and 162) are applicable to all microbial pesticides, both naturally occurring and otherwise.

The agency believes that these requirements are adequate for the assessment of indigenous microbial pesticides, and provide a basis for evaluating genetically engineered and nonindigenous microbial pesticides as well. However, the Agency believes that additional data and information, determined on a case-by-case basis, may be necessary to evaluate some properties of genetically engineered and nonindigenous microbial pesticides. Part 158 explicitly provides the necessary flexibility to require additional data (§ 158.65) as well as the flexibility to waive data requirements that are not applicable (§ 158.45).

c. *Additional requirements for genetically engineered and nonindigenous microbial pesticides.* Any additional data requirements will be determined on a case-by-case basis depending on the particular microorganism, its parent microorganisms, its native habitat, the pesticide use pattern, and the manner and extent to which the microorganism may have been engineered. These additional requirements could include:

(1) Description of the natural habitat of the microorganism or its parental strains, including information on natural predators, parasites, and competitors.

(2) Information on relative ability to survive and increase in number or biomass as compared to the parental strains.

(3) Selected environmental fate tests from 40 CFR 158.170.

(4) Additional toxicology tests from 40 CFR 158.170.

(5) If the microorganism is genetically altered, then information on the genetic modification techniques used, the identity of inserted gene segment(s) (base sequence data or restriction enzyme map of the gene), the control region of the gene(s), a description of the new traits or characteristics that are intended to be expressed, and tests to evaluate genetic stability and exchange, may be required as specified previously at Unit II.D.1.b above.

d. *Review process for genetically engineered and nonindigenous microbial pesticides.* EUP applications will be reviewed in compliance with the EUP regulations under 40 CFR Part 172. The registration, reregistration, and classification procedures of 40 CFR Part 162 will be followed for registration applications. The review process will contain the same major elements as those outlined previously for small-scale field testing notifications (see Unit II.D.1.c). Briefly, this process involves scientific review and risk assessment by EPA scientists and, if appropriate, review and comment from other Federal agencies and independent expert consultants.

Once the supporting data have been submitted, EPA has up to 120 days to review an EUP application and determine whether to grant a permit. Past experience indicates that the registration process for a new microbial pesticide may vary from 9 months to several years depending upon the particular product, its use pattern, and the completeness of the registration package submitted to EPA.

Both the EUP and registration process may provide an opportunity for public comment. For example, § 172.11 of the EUP regulations specifies that if an application may be of regional or national significance the Agency will announce receipt of the application in the Federal Register. The announcement is accompanied by a description of the experimental program and public comments are solicited. Similarly, § 162.6 of the registration regulations specifies that if a registration application relates to a new active ingredient or a new use, notice of receipt of that application shall be published in the Federal Register with a request for public comment. Information on the submission is made available for public inspection.

EPA has several regulatory options for responding to either an EUP or registration application. For example, after completing its review, the Agency may determine that the field test or registration poses no unreasonable risks to humans or the environment and may grant the application. Alternatively, EPA may conclude that some additional information or data are needed to assess the potential risks adequately. In this case, the application would be asked to provide the necessary data before EPA would decide whether to grant the application. In other cases, the Agency may impose additional limitations or restrictions on the field test or registration to address a potential risk. Finally, EPA will deny those applications where it has determined that it has all the necessary data to complete a risk assessment and that the field test or registration would pose an unreasonable risk to humans or the environment, even if additional limits or restrictions are imposed.

III. Applicability of the Toxic Substances Control Act (TSCA) to Microbial Products

A. Overview of This Unit

As discussed in the December 84 notice (49 FR 50886), EPA will review certain microorganisms and uses of microorganisms under TSCA. Microorganisms and their DNA molecules are "chemical substances" under section 3 of TSCA, and thus are subject to all the provisions of TSCA, except to the extent they are manufactured, processed, or distributed in commerce for use as pesticides, foods, food additives, drugs, cosmetics, and medical devices. For purposes of analysis and convenience of administering TSCA, EPA has chosen to focus on the microorganism as the "chemical substance."

This unit explains the statutory requirements of TSCA as they apply to microorganisms. It begins by describing which microorganisms are within the scope of TSCA and which are not. Following that are units describing five categories of microorganisms or uses of microorganisms that are or will be subject to reporting requirements under TSCA.

B. Scope of TSCA

Many organisms are not subject to TSCA requirements because of statutory exemptions; others will be exempt from certain TSCA requirements as a matter of regulatory policy. In general, the use of a microorganism determines whether it is subject to TSCA or to other laws.

Many of the comments received by OTS indicated

misunderstandings of TSCA's scope. Therefore, those organisms which are and are not subject to TSCA are described in this Unit.

1. *Organisms not subject to TSCA* — a. *Microbes used as foods, food additives, drugs, cosmetics, medical devices, and pesticides.* Microorganisms are sometimes used directly as foods, food additives, drugs (including both human and animal vaccines), cosmetics, medical devices, and pesticides. When microorganisms are used for these purposes, they are explicitly excluded from TSCA and from the policies described in the TSCA portions of this notice (TSCA section 3(2)(B), 15 U.S.C. 2602(2)(B)). Microorganisms that are used as foods, food additives, drugs, cosmetics, medical devices, and pesticides are regulated by the Food and Drug Administration (FDA), USDA, or the EPA Office of Pesticide Programs. Applicable requirements for pesticides are described in Unit II of this notice. Requirements for foods, food additives, drugs, cosmetics, and medical devices are described in the FDA and USDA notices in this Federal Register.

b. *Microbes used to produce foods, food additives, drugs, cosmetics, and medical devices.* In addition to being used themselves for food, drug, and other purposes, microorganisms are often used to produce chemicals that are in turn used for such purposes. For reasons explained in the December 84 notice, microorganisms will not be reviewed under TSCA when used to produce foods, food additives, drugs (including vaccines), cosmetics, or medical devices. Further information on these uses may be found in the FDA and USDA notices in this Federal Register.

Microorganisms used in the production of chemical end products other than foods, food additives, drugs (including vaccines), cosmetics, and medical devices are subject to

TSCA. They are described in Unit III.B.3 below.

2. *Plants and animals not subject to these policies.* Plants and animals are not subject to the TECA policies in this notice, either as whole organisms or as *in vitro* cultures for the reasons set forth in the December 84 Notice. (Definitions of plants and animals for regulatory purposes are provided in Unit IV of this EPA notice.) There are two exceptions to this general rule. First, if plant or animal gene segments are intentionally incorporated into microorganisms, the microorganisms that contain those plant or animal genes may be subject to TSCA, depending on how they are used (see Units III.B. 1 and 3). Second, a chemical extracted from a plant or animal may be subject to TSCA, again depending on how it is used. The USDA and FDA notices in this Federal Register contain information about regulations that apply to plants and animals.

3. *Organisms subject to TSCA — microorganisms used for purposes not excluded by law.* With the exceptions described above, all microorganisms produced for environmental, industrial, or consumer uses are potentially regulable under TSCA. It is not possible to list all the applications that could be subject to TSCA because many are yet to be developed. Some of the microorganisms that are expected in the near future and that would be subject to TSCA include microorganisms used in conversion of biomass for energy, pollutant degradation, enhanced oil recovery, metal extraction and concentration, and certain non-food and non-pesticidal agricultural applications, such as nitrogen fixation.

Microorganisms used in the production of a chemical end product will be subject to TSCA if the end product is any chemical substance used for a purpose other than as a food, food additive, drug, cosmetic, or medical device. For example, microorganisms are subject to TSCA if they are used in the production of pesticides, fuels, solvents, dyes, cleansing agents, etc. TSCA jurisdiction over such microorganisms, which may be used entirely in closed manufacturing systems, is consistent with TSCA coverage of conventional chemicals. For example, chemical intermediates — even those used in closed systems — fall under TSCA authority and are subject to PMN requirements if new (40 CFR Part 720). Similarly, as described in Unit III.C. 1 of this notice, "new" microorganisms used in chemical production are subject to PMN requirements.

4. *Chemicals produced by microorganisms — Status under TSCA.* Although the purpose of this notice is to provide information on the applicability of TSCA to microorganisms, some readers may wish to obtain information on requirements that apply to chemicals produced by microorganisms. For example, various proteins and polysaccharide gums are produced by microorganisms and may be subject to TSCA, depending on how they are used (see Unit III.B.1). These chemicals produced by microorganisms are subject to the same requirements and procedures as chemicals produced by other means. Any special concerns pertaining to the microbial production method, such as the possibility of contaminants, will be considered during the review of the microorganisms used in producing the chemicals. This approach is explained in the December 84 notice at page 50890.

C. Specific Requirements Under TSCA

The fact that a microorganism is potentially subject to TSCA does not necessarily mean that it will be regulated under TSCA. The rest of this unit explains the specific provisions that apply or will apply to various types of microorganisms falling within TSCA's jurisdiction.

In overview, microorganisms are (or will be) subject to TSCA requirements in the following manner:

As of the date of this notice, microorganisms that are subject to TSCA and contain genetic material from dissimilar source organisms (i.e., organisms from different genera) are subject to PMN requirements.

Microorganisms other than intergeneric combinations that are subject to TSCA and are pathogenic or contain genetic material from pathogens, will in the future, if released into the environment, be subject to "significant new use" reporting requirements under TSCA section 5(a)(2). One exception is that agricultural uses of such microorganisms will be reviewed by USDA rather than EPA. EPA expects voluntary notification to begin immediately for uses that will be subject to significant new use reporting requirements.

The research and development exemption from PMN and significant new use notification requirements will be amended so that it no longer applies to microorganisms released to the environment. EPA expects voluntary notification of such uses to begin immediately.

EPA will issue a rule requiring manufacturers and importers to submit general information on environmental uses of microorganisms that are subject to TSCA but not otherwise subject to notification requirements, so that EPA can monitor environmental releases.

All manufacturers, processors, and distributors of microorganisms subject to TSCA are reminded of the requirement to report any information on substantial risks under TSCA section 8(e).

EPA is considering initiating rulemaking that would exempt from PMN requirements inter-generic microorganisms used solely in contained systems and never intentionally released to the environment.

1. *Premanufacture notification requirements* — a. *Overview.* EPA has determined that any microorganisms that are subject to TSCA (described in Unit III.B), and that through deliberate human intervention contain genetic material from dissimilar source organisms, are "new" and therefore subject to PMN requirements of TSCA. This interpretation is effective as of the date of publication of this notice.

Organisms are considered dissimilar for the purposes of this policy if they are from different genera. In the case of chemically synthesized genes, the Agency will follow the same principle, as clarified below in Unit IV. Detailed guidance on how to determine if organisms are from different genera is also provided in Unit IV.

The agency is excluding certain inter-generic combinations from PMN requirements, i.e., those inter-generic combinations in which the genetic material added to the recipient microorganism consists only of well-characterized, non-coding regulatory regions (see Unit IV). The resulting microorganisms do not possess new combinations of traits but rather exhibit quantitative changes in preexisting traits.

EPA is leaving unanswered, for now, the question of whether microorganisms containing genetic material from other microorganisms in the same genus (i.e., products of deliberate intra-generic combinations) and those which are developed from a single source microorganism (e.g., products of undirected mutagenesis, microorganisms with deletions) should also be considered "new." In the future, it is possible that EPA will decide that such microorganisms are "new," but for now they are not subject to PMN requirements.

b. *Background.* For purposes of administering TSCA, EPA must decide what constitutes a "new" microorganism which is subject to PMN requirements. As mentioned in the introduction to the EPA portion of this notice, EPA originally proposed a "process-based" approach to determining whether a microorganism is new. This approach stated that a microorganism would be considered new if significant human intervention had been used in developing it. For example, microorganisms altered by certain techniques — such as recombinant DNA and cell fusion — were presumed to be new because they involved significant human intervention. The question of which other techniques should be considered to produced new microorganisms was left open and comments were solicited.

After reviewing the comments, EPA considered a number of alternative ways to define "new" microorganisms. These are described in the "Response to Comments" document available as background to this Federal Register notice. In choosing among the alternatives, EPA carefully considered the TSCA mandate to review "new" substances. The Agency also considered related issues, for example, how well the options approximated risk (there was uncertainty with all the options in this respect) and how readily they could be implemented and enforced.

c. *Rationale.* Having reviewed the TSCA section 5 PMN requirements, the PMN regulations, the public comments, and the current state of science regarding genetic engineering, EPA has concluded that microorganisms resulting from intentional, inter-generic combinations of genetic material, except those in which the transferred material is only a well-characterized, non-coding regulatory region, constitute new microorganisms for purposes of PMN reporting. The reasons for this are set forth below.

First, the Agency considered the regulatory precedents established in compiling the inventory of existing chemical substances under section 8(b) of TSCA. Any chemical substance not on this inventory is "new" under section 5(a) of TSCA and is therefore subject to PMN requirements. Naturally occurring substances and substances derived from nature with limited human intervention are not explicitly listed on the inventory but are considered implicitly to be on it, and thus are not "new" (see 40 CFR 710.4(b)). A more detailed explanation of the TSCA inventory and related issues is found in the December 84 notice at pages 50887-50888.

Second, the Agency evaluated these regulatory precedents in the light of scientific knowledge about genetic engineering and microorganisms found in nature. On this basis, EPA concluded that microorganisms found in nature and developed without any deliberate combination of genetic material are not new, because they occur naturally and are derived through limited human intervention. Furthermore, from a scientific standpoint, these microorganisms have a very low probability of exhibiting new combinations of traits. Therefore, the Agency considers that from a legal and scientific standpoint they must be considered naturally occurring (not new). Because such microorganisms are naturally occurring, they are, as explained above, implicitly listed on the TSCA chemical substances inventory and not subject to PMN requirements.

Third, where genetic material has been combined among source organisms from different genera (inter-generic), the resulting microorganisms should be considered "new" because of the degree of human intervention involved, the significant likelihood of creating new combinations of traits, and the greater uncertainty regarding the potential risks of such microorganisms. However, transfer of genetic material consisting solely of well-characterized, non-coding regulatory regions is a special case. Where only regulatory material is transferred, no distinctly new combinations of traits are introduced; instead, existing traits in the receiving microorganisms are amplified or changed quantitatively. For this reason, EPA believes that microorganisms formed only through inter-generic transfer of well-characterized, non-coding regulatory regions should not be considered "new" under section 5 of TSCA. This is reflected in the definition of "inter-generic" found in Unit IV.A.

It is possible to argue that some microorganisms formed through intra-generic combinations are products of significant human intervention and may exhibit new combinations of traits, and therefore that they should also be considered new. However, the Agency at this time believes that it is appropriate to exclude such microorganisms form its definition of "new" because distinctly new combinations of traits are unlikely to occur through transfers of genetic material among closely related organisms, because transfers among closely related organisms are more likely to occur in nature, and because the current state of taxonomy with regard to species designations is sufficiently unstable that it makes it difficult to include such microorganisms in a definition of "new" (the rationale is found in Unit I.D.3.a). As explained previously, however, the Agency will continue to review the status of such microorganisms and may, in the future, determine that certain combinations among similar organisms should be considered new.

In summary, EPA considers microorganisms deliberately formed to contain genetic material from different genera to be new, except where only well-characterized, non-coding regulatory regions are transferred. Conversely, intra-generic and nonengineered microbes are considered naturally occurring. These conclusions are based on the TSCA section

5 mandate to review "new" substances, and they also reflect a number of scientific considerations. Among these are (1) the Agency's concern that microorganisms formed with genetic material from different genera warrant regulatory review, because of the inherent uncertainty about the characteristics and behavior of such microorganisms, (2) the observation that microorganisms from different genera are less likely to exchange genetic material in nature than microorganisms that are more closely related, (3) the regulatory precedent that significant human intervention creates new substances for purposes of PMN under TSCA section 5, and (4) the necessity of having a definition of "new" that can be readily interpreted and enforced given the current state of science. These scientific and legal issues are more fully described in Unit IV.A.

d. *How to comply with the PMN requirements for new microorganisms.* The following requirements apply to "new" microorganisms produced for uses subject to TSCA authority (see Unit III.B.1 and 3). Detailed criteria for determining whether a microbe meets the definition of "new" microorganism (i.e., whether it contains genetic material from organisms from different genera) may be found in Unit IV.A.

Certain PMN policies in this notice are immediately effective. As of the date of publication of this notice, microorganisms that are being manufactured or imported for any TSCA commercial purposes other than research and development (R&D) are subject to PMN requirements 90 days prior to manufacture or import. This requirement applies to both contained and environmental uses that have gone beyond R&D. The requirement is based on the current provisions of 40 CFR Part 720. The definition of R&D under these regulations is clarified in the Federal Register of April 22, 1986 (51 FR 15096).

In addition, new microorganisms that are being manufactured or imported for R&D that involves environmental release will have to be reported to EPA at least 90 days before such activities begin. This policy will be implemented through amendments to 40 CFR 720 (explained fully in Unit II.C.3); in the meantime, persons manufacturing or importing new microorganisms for R&D activities involving environmental release are expected to comply with this policy voluntarily.

EPA believes that these are no manufacturers who are presently beyond the research and development stage with new microorganisms subject to TSCA. However, if any companies are now engaged in such activities, they should contact EPA and determine whether a PMN is necessary. If a company in this position contacts EPA promptly, it will not be considered out of compliance with policy. Further information on TSCA PMN requirements may be obtained from the TSCA Assistance Office (address provided at the beginning of the EPA portion of this notice).

(1) *How to know if a microorganism is subject to PMN.* As stated above, all microorganisms containing deliberate combinations of genetic material from organisms from different genera are new and subject to PMN. An exception to this policy is an inter-generic combination in which the genetic material added to the recipient microorganism consists only of well-characterized, non-coding regulatory regions. Unit IV.A of this notice contains detailed guidance that manufacturers should use to determine if their microorganisms meet this definition.

Submitters should consult the Agency if they have any questions about how to determine if a microorganism contains genetic material from different genera.

(2) *PMN exemptions.* EPA considers it a priority to exempt from PMN requirements new microorganisms that can be shown to meet the findings for exemption under TSCA section 5(h)(4). Further information on exemptions the Agency is considering may be found in Unit III.C.6 of this notice.

(3) *Submitting the PMN.* EPA expects manufacturers and importers to contact EPA well in advance of PMN submission, to allow sufficient time for prenotice consultation. These consultations will help the Agency and the submitter anticipate potential problems and expedite the review.

Information regarding new microorganisms should not be submitted on the standard PMN form, as this form is not applicable to microbial products. Instead, EPA and the submitter will discuss the level and types of information appropriate for the notice during prenotice consultations. The general kinds of information EPA expects to see in most submissions for microorganisms are described in the next unit below.

(4) *What information to submit.* Section 5(d)(1)(A) of TSCA specifies the information PMN submitters must provide in their notices, including information on production, workplace exposure, and release. In addition, under section 5(d)(1)(B) submitters must provide all test data related to the health and environmental effects of the new chemical substance in their possession or control. For more information on PMN requirements, persons should consult EPA's PMN rule (40 CFR Part 720).

In general, information to assess a substance's potential risk should be developed in a step-wide fashion. PMN submitters should begin with published literature on the source organisms, then move through laboratory, microcosm, growth chamber, and/or greenhouse studies that simulate as closely as possible the conditions of the eventual use or environmental application.

The remainder of this unit describes the types of information EPA expects submitters to provide in PMNs on new microorganisms.

(a) *Identifying the microorganism.* PMN submitters must provide information that identifies microorganisms well enough to be listed on the TSCA chemical substance inventory. If the identity and/or use of the microorganism are claimed as confidential business information by the submitter, the PMN must also include a generic description of these items so that the information can be published in the Federal Register. Confidential submissions will be considered incomplete unless this generic information is included (see 40 CFR 720.65, 720.85, and 720.87).

Once a new microorganism is actually manufactured or imported, it will be listed on the inventory and will be no longer subject to PMN requirements. (See 40 CFR 720.102 concerning submission of a Notice of Commencement of Manufacture or Import.) EPA proposed an approach to inventory listings in a background document to the December 84 notice. The Agency received very few comments on this document, but those who commented stated that a general method for listing all microbes does not seem possible at this time. The Agency agrees and therefore intends to list microorganisms on the inventory on a case-by-case basis while developing more general procedures for different classes of microorganisms, and gaining experience that will help in developing standard listings. For now, the inventory definition will usually include the genus and species designations of source organisms and of the microorganism being reported, and other relevant phenotypic information such as nutritional and substrate requirements, proteins expressed, primary characteristics for which the microbe was engineered, and characteristics that are a typical for the species.

To identify the microorganism, EPA is likely to require information on:

i. Source organisms (e.g., taxonomy, source, reproductive cycle, and capacity for genetic transfer).
ii. Methods used to manipulate source organisms genetically to obtain the resulting product (e.g., source and function of genetic material to be combined; description of methods for vector construction and introduction, fusion of cells, injection of DNA, etc.).
iii. The special functions obtained (e.g., new traits intended to be expressed; selection method; nature and amount of source genetic material remaining in the product microorganism; genetic stability of new trait).

(b) *Risk assessment information.* Data required for conducting the risk assessment will vary according to the specifics of each case, but in general will fall into several major categories: Information on exposure, environmental fate, and human health and environmental effects.

If the microorganisms will be produced in enclosed, commercial-scale facilities, or used solely in physically contained systems, the notice should include the following information:

i. Production processes (e.g., culture conditions and requirements; sites, methods, and amounts of manufacture, processing, storage, and shipment; volume, composition, and disposal of wastes).

ii. Workplace exposure and worker practices (e.g., potential for exposure, worker protection practices, and equipment).

iii. Containment and possible releases (e.g., potential sources and characteristics of releases, physical containment methods, emergency backup systems, monitoring, and detection methods in event of a release).

In the case of small-scale field tests and other environmental releases, EPA expects that the submitter will provide information on:

(A) Purpose and intended effect of application.

(B) Site of application and surroundings, including geographic, physical, chemical, and biological features.

(C) Numbers of microorganisms and methods of application.

(D) Containment and mitigation measures (e.g., procedures in event of accidental release, for emergency termination of the application, and to reduce dispersal beyond the site).

(E) Monitoring (e.g., detection procedures including their limits, sampling procedures).

For field tests and other environmental releases, data on environmental fate and effects will be essential. In such cases, manufacturers should assume, in the absence of data to the contrary, that the microorganisms may present a risk because of their potential to reproduce and exhibit new traits. Therefore, EPA will expect manufacturers to provide test and other data demonstrating the microorganisms' safety. These data should include:

(i) General background information on the source organism (e.g., habitat and geographic distribution, interactions with other organisms, involvement in biological cycling processes, potential for genetic exchange in nature).

(ii) Test data on the new microorganism itself, indicating its potential for survival, replication, dissemination, and genetic exchange with other organisms.

For further guidance, manufacturers should refer to the "Proposed Points to Consider for Environmental Testing of Microorganisms" developed by the National Institutes of Health Recombinant DNA Advisory Committee Working Group on Release into the Environment (Ref. 11). This document is particularly useful in developing data and information for submissions on small-scale field tests. While some points in this document relate solely to recombinant DNA techniques, most of the considerations are relevant to environmental tests of microorganisms regardless of the techniques involved in their production.

In addition, the Agency has prepared a more detailed guidance document entitled "Points to Consider in the Preparation and Submission of PMNs for Microorganisms." This document provides guidance on both environmental and industrial applications of microorganisms and is available from the TSCA Assistance Office (see address at the beginning of this notice).

At least three other documents will be useful to submitters. These are the "EPA Pesticide Assessment Guidelines: Subdivision M — Biorational Pesticides" (Ref. 20), a National Science Foundation report titled "The Suitability and Applicability of Risk Assessment Methods for Environmental Applications of Biotechnology" (Ref. 3), and a report by the Cornell Ecosystems Research Center titled "Potential Impacts of Environmental Release of Biotechnology Products: Assessment, Regulation, and Research Needs (Ref. 9).

e. *The PMN review,* All reviews of microorganisms will follow established administrative steps that are the same for all substances subject to PMN review. First, within 5 days of receiving the PMN, EPA will issue an announcement in the Federal Register describing the submission. The anouncement will include information on the identity of the new microorganism, the type of use, occupational exposure, production volume, a summary of test data submitted in the notice, and the submitter's identity. It will have confidential business information deleted according to the manufacturer's instructions, although EPA will strongly encourage manufacturers to release as much information as possible. If identity and use are claimed confidential, the Agency will include a generic description provided by the submitter. EPA will have 90 days to review the PMN (extendable to 180 days), during which time the microorganism cannot be manufactured or processed for purposes other than research and development. Within the review period, the Agency may take action under section 5(e) of TSCA to prohibit or limit the activities, pending receipt of more data, or under section 5(f) or 6 to prohibit or limit the activities if there is sufficient information to make an unreasonable risk finding. Alternatively, EPA may take no action. In this case, manufacture and use may begin without restriction.

(1) *Case-by-case assessments.* Because of the very recent development of genetically engineered microorganisms for environmental use, there is little direct experience for conducting risk assessments on environmental releases of engineered microorganisms. In the absence of such experience, the Agency will conduct case-by-case reviews

by using information from various scientific disciplines and by directly considering the features of specific genetically engineered microorganisms and their uses.

Many existing risk assessment approaches that are used for nonengineered microorganisms will contribute to the analysis of risks of engineered microbes in the environment. Some of these will be adopted with few if any changes, while others will require modifications to address special problems.

EPA believes that standardized protocols and procedures should be gradually blended with the case-by-case approach. As experience is gained, increasingly detailed guidance on routine testing and procedures can and will be developed.

(2) *Use of experts.* Expert judgment will be critical in determining information needs, evaluating protocols for testing, and reviewing potential risks. Because of the range of expertise that may be required in any given case, EPA intends to supplement its staff expertise by using experts from other government agencies, academia, and other independent sources. Persons will be specifically chosen for their knowledge and experience with organisms and uses related to the PMN under review.

As announced in the December 84 notice (and further described in Unit I of this notice), EPA is forming a biotechnology Science Advisory Committee to provide scientific advice and promote consistent review procedures.

Many academic experts may have financial or contractual relationships with biotechnology companies. Using non-Agency experts to assist in PMN reviews may therefore raise two potentially sensitive issues: Conflicts of interest and access by non-Agency experts to confidential business information. To address these issues, the EPA Office of

Toxic Substances has developed special procedures to ensure that scientists contributing to biotechnology PMN reviews will not have conflicts of interest, and will have the necessary access to CBI to review the PMN without compromising trade secrets or violating TSCA CBI procedures. A document describing these procedures will be placed in the public record for this policy statement.

(3) *Major parts of the review process.* As stated earlier, EPA expects persons developing biotechnology products to engage in prenotice consultations with the Agency. During these discussions, EPA and the consulting company can identify preliminary concerns by considering the source organisms and intended uses of the microorganism subject to PMN. Significant time may be saved later in the PMN process if these concerns are addressed before the PMN is submitted.

Once the PMN is submitted, EPA will develop hazard and exposure assessments based on information submitted in the PMN, other available information, and consultation with non-Agency experts. Reviewers will consider the types of issues and questions described here and in the various guidance documents on risk assessments for microorganisms. As appropriate, they may also consult with external scientific experts, and their analyses may be peer reviewed by the Agency's biotechnology Science Advisory Committee.

As a risk/benefit statute, TSCA requires that benefits be estimated and considered in judging whether the risk may be unreasonable. While the risk assessments are being developed, Agency economists will estimate the benefits of the product based on information from the submitter, independent economic research, and consultation with non-Agency experts.

Finally, EPA staff will prepare a summary of the risks and benefits to use in reaching regulatory decisions.

(4) *Public involvement in the review.* EPA will issue for publication a section 5(d)(2) notice after receipt of a PMN for a new microorganism. EPA will also maintain a copy of the PMN, from which CBI has been deleted, in the OTS Public Information Office at the address listed in Unit VI of the EPA notice. EPA will welcome comments from interested members of the public on the PMN. The public is generally given 30 days to comment on a PMN after publication of the section 5(d)(2) notice.

In addition to the normal procedures for public comment on PMNs, EPA intends that meetings of its biotechnology Science Advisory Committee will be open to the public, although certain portions of meetings may have to be closed to discuss CBI. EPA also intends to charter its committee to include representatives from the lay public. These features will help to ensure that the public has access to information about EPA biotechnology policies and decisions.

(5) *Possible regulatory decisions.* The Agency may come to one of three decisions at the conclusion of a particular PMN review: (a) There is sufficient information to determine that the risks are reasonable, (b) there is sufficient information to determine that the risks are unreasonable, or (c) there is insufficient information to make a reasoned evaluation of risk, and the substance may present an unreasonable risk or there may be significant or substantial exposure to it.

Where the first decision is made, the Agency will notify the PMN submitter that the manufacture and use may proceed without restriction. In any event, unless the Agency notifies the company to the contrary before the end of the 90-day review period (with a possible 90-day extension), the submitter may begin to manufacture and use the organism.

A decision that risks will be unreasonable leads to two regulatory options. The Agency may require measures to reduce the risks to an acceptable level as a condition of manufacture and use. Alternatively, the Agency may prohibit manufacture or use of the microorganism if there are no alternatives available or practical to reduce the risk sufficiently. Such actions can be taken under TSCA section 5(f).

If the information submitted with the PMN is insufficient for a reasoned evaluation, and EPA finds that the microorganism may present an unreasonable risk or that there may be significant or substantial human exposure to it, or substantial environmental release, EPA may, under TSCA section 5(e), limit or prohibit the manufacture or use of the microorganism until sufficient data are submitted to the Agency to evaluate the risks.

2. *Significant new uses of microorganisms* — a. *Overview.* EPA intends to supplement its PMN requirements by requiring persons to notify the Agency before they introduce pathogenic microorganisms (including microorganisms containing genetic material from pathogens) into the environment. Notification will be required for new environmental applications of genetically engineered pathogens prior to their release in any amounts into the environment, while notification for nonengineered pathogens will be required at a somewhat later stage, prior to their introduction on more than 10 acres of land (or some equivalent measurement standard in cases where acreage is not applicable, e.g. aquatic uses). If a pathogen used for agricultural purposes is subject to USDA review, it will not be subject to this policy. Applicable definitions may be found in Unit IV.

EPA intends to implement these notification requirements through a significant new use rule (SNUR) under TSCA section 5(a)(2). The public will have the opportunity to comment on the proposed rule, including its scope and possible categories that could be excluded from coverage.

Until the rule is final, EPA expects persons introducing pathogens into the environment for non-agricultural new uses to report to EPA voluntarily. In the unlikely event that an imminent hazard would arise during this interim period, the Agency could use its authority under section 7 of TSCA to immediately limit or prohibit the manufacture, processing, distribution in commerce, use, or disposal of the hazardous product.

b. *SNUR background.* Section 5(a)(2) of TSCA (15 U.S.C. 2604(a)(2)) authorizes EPA to determine that a use of a chemical substance is a significant new use. The Agency must make this determination by rule, after consideration of all relevant factors, including those listed in section 5(a)(2). Once EPA determines that a use of a chemical substance is a significant new use, section 5(a)(1)(B) of TSCA requires persons to submit a notice to EPA at least 90 days before they manufacture, import, or process the substance for that use.

Persons subject to a SNUR must comply with most of the same notice requirements and regulatory procedures as submitters of PMNs under section 5(a) of TSCA. EPA's review procedures and regulatory authority are the same for SNUR notices as for PMNs. However, if EPA does not take action on a SNUR notice, section 5(g) of TSCA requires the Agency to explain in the Federal Register its reasons for not taking action. Procedures and requirements for PMN review are described above in Unit III.C.1.

c. *SNUR rationale.* As explained in the December 84 notice, EPA recognizes that any approach to defining "new" microorganisms, including the one described in Unit III.C.1, excludes some types of microorganisms from PMN review and therefore may not address some significant potential risks. EPA believes there is one currently identifiable category of microorganisms that is not being treated as "new" under TSCA at this time but that should be reviewed before environmental release. That category includes pathogens and microorganisms that contain genetic material from pathogens (henceforth, both are referred to collectively as "pathogens"). As explained in more detail in Unit I, the Agency believes it is necessary to review pathogens released to the environment because of their ability to cause disease in microbes, plants, animals, and humans.

EPA intends to take a slightly different regulatory approach with nonengineered pathogens. The Agency will not require SNUR reporting on the use of nonengineered pathogens until they are to be used on more than 10 acres of land, or some equivalent standard (to be determined) for uses where acreage is an inappropriate standard (e.g. aquatic or subterranean uses). The reason for this exception is explained in Unit I.D., "Rationale for Approach."

To avoid duplicative requirements with USDA, EPA will exclude pathogens used solely for agricultural purposes from the scope of its SNUR, USDA permits to use such microorganisms are mandatory, while EPA review would be discretionary because these are not "new" microorganisms. However, new environmental applications of pathogens for non-agricultural purposes will be subject to EPA review as significant new uses, and will in some cases also be subject to USDA oversight (if they are plant or animal pests under the USDA definition). In such cases, USDA's review will primarily be for the purpose of detecting potential adverse agricultural effects, while EPA's review will focus on the potential non-agricultural impacts. See Unit I.E. for an explanation of how the agencies will work together to coordinate their review.

EPA is considering whether it should also include provisions in the SNUR requiring notification prior to small-scale releases or commercial uses of other categories of microorganisms besides pathogens. For example, some people have expressed concern over nonindigenous microorganisms, and others have expressed concern over microorganisms that degrade structural components of nature such as lignin and cellulose. Members of neither category are subject to PMN when the microorganisms involved are naturally occurring or intra-generic (not new), and they would not be subject to the provisions for pathogens described above. However, they may present certain risks because they are new to the environment in which they are used or because of their degradative capabilities. The literature contains much documentation of the adverse effects that have occasionally been caused by nonindigenous microorganisms such as the chestnut blight fungus and Dutch Elm disease fungus. There is, on the other hand, very little known about many degradative microorganisms and their potential for adverse effects. The Agency will request comments on these concerns when it issues its proposed SNUR.

d. *Guidelines for voluntary compliance.* The SNUR that EPA will propose will describe, in detail, the parsons who will be subject to the rule and the microorganisms and activities for which significant new use reporting will be required. In the meantime, EPA strongly encourages persons who are planning to manufacture, import, or process pathogenic microorganisms for non-agricultural, new environmental uses, except those used solely for agricultural purposes, to report their activities to the Agency and to provide information similar to that required for a PMN for a new microorganism.

For purposes of voluntary reporting, persons may use the following definitions and assumptions. These guidelines may be changed in the proposed and final forms of the SNUR.

(1) *How to know if a use would be considered a significant new use.* For purposes of voluntary reporting, the Agency encourages people to be as comprehensive as possible and to consider that any new, non-agricultural release of a pathogen to the environment is appropriate to report. "Environmental release" is defined in Unit IV.D, this definition should be used in the interim until the SNUR is final. Cases that may not be entirely clear, e.g., use in waste water treatment plants and use in mines or oil wells, should be reported until the Agency provides further guidance.

Many microorganisms that are pathogens or that contain genetic material from pathogens are being used in the environment already. For example, specific naturally occurring pathogens are used for waste treatment purposes and are tested in non-contained experiments. These applications of these specific microorganisms cannot be considered significant "new" uses because they are ongoing. Therefore, persons now using pathogens in environmental applications will not be expected to notify the Agency of such uses of these pathogens, except for informational purposes (see Unit III.C.4).

In developing the proposed and final rule, the Agency will have to determine exactly which types of uses should be considered significant new uses, taking into account that the purpose of the rule is to ensure the Agency has the opportunity to review releases of pathogens that could entail significant exposure or risk to the environment or the public. Considerations relating to the appropriate scope of the rule will be discussed in the proposed SNUR, and the public will be invited to comment.

(2) *How to know if a microorganism is a pathogen.* Unit IV.B of this notice contains the definition of "pathogen" that the Agency will use for purposes of administering TSCA and FIFRA, and provides guidance on how to determine if a microorganism is a pathogen.

(3) *How to know if a microorganism is genetically engineered.* As discussed in Unit III.C.2.c, EPA will not require nonengineered pathogens to be reported until they are used on more than 10 acres of land (or some equivalent standard, not yet determined, for uses where acreage is an inappropriate standard). For now, a pathogen should be considered nonengineered if there has been no deliberate attempt to promote genetic changes. Any human intervention beyond removal from the environment and selection for the desired variant populations should be considered to result in an engineered microorganism.

(4) *Submitting the significant new use notice.* Persons subject to the SNUR will have to notify the Agency at least 90 days prior to any new, non-agricultural use involving environmental release of engineered pathogens. The Agency will treat nonengineered pathogens slightly differently; producers of nonengineered pathogens will be subject to significant new use notification 90 days prior to new uses involving environmental applications on more than 10 acres of land. Significant new use notifications for microorganisms should contain the same general types of information as PMN submissions for microorganisms. In all cases, SNUR notice submitters should initiate prenotice consultations

with EPA well in advance of the actual submission, to expedite the Agency's review of the notice.

e. *Significant new use notice review.* EPA reviews of significant new uses of microorganisms will be conducted in a fashion similar to PMN reviews of microorganisms. The review must be completed in 90 days, extendable for good cause to 180 days. In conducting the review, EPA will use Agency and non-Agency scientists selected for their expertise on issues relevant to the specific cases.

The Agency recognizes that various environmental uses of different types of pathogens pose very different levels of potential risk to human health and the environment. For example, risks should generally be lower when pathogens are applied in areas distant from host organisms; the manufacturer has used nonpathogenic strains of a pathogenic species; transferred genes are for a trait not directly involved in pathogenicity; the pathogenic source organisms have very narrow host ranges; and pathogenic genes have been deleted.

Because it recognizes these variations in risk, the Agency expects to subject some pathogenic microorganisms to more rigorous regulatory oversight than others.

3. *Research and development (R&D) exemption* — a. *Overview.* TSCA section 5(h)(3) exempts from PMN and SNUR notification requirements chemical substances manufactured in small quantities solely for R&D. However, to ensure adequate review prior to environmental release, EPA intends to require persons developing "new" microorganisms and certain engineered pathogens to notify EPA prior to any research involving environmental release. This will be accomplished by amending the PMN rule (and possibly the general SNUR rules in 40 CFR Part 721) to

specify that field testing of microorganisms does not fall within the definition of "small quantities" for R&D. Until the necessary rule changes implementing this policy are final, EPA expects submitters to comply with this policy voluntarily. Notice submitters are advised to consult the Agency if they are unsure whether a particular test is subject.

b. *Background.* As explained in the December 84 notice (at page 50891), section 5(h)(3) of TSCA exempts from PMN requirements new chemical substances produced "only in small quantities solely for purposes of research and development." ("Small quantities" must be defined by rule.) The same exemption applies to substances produced for significant new uses. If this exemption as now defined were applied to living microorganisms, many microorganisms would go unreviewed by EPA until perhaps years after their initial testing in the environment. Because microorganisms can reproduce in the environment and have the potential to exhibit new traits, this has raised the question of whether these field tests for R&D purposes could present significant risks that would go unreviewed.

Because of this concern, an important issue for EPA in implementing the biotechnology program has been whether to alter the R&D exemption of TSCA section 5 notice requirements in the case of living microorganisms. EPA requested and received substantial public comments on this issue, which it considered carefully in developing this policy. The comments and EPA's response to them are described in the EPA "Response to Comments" document, available as part of the public record of this EPA notice.

c. *Rationale.* The PMN rule definition of "small quantities" for R&D has been appropriate for most chemicals subject to TSCA because of the assumption that chemical

R&D generally involves limited exposure and therefore limited risk. In the case of field tests involving living microorganisms, this assumption will not always apply. Microorganisms that survive may reproduce, potentially leading to significant exposure and risks. Because of their ability to reproduce and therefore increase beyond the amount originally released, living microorganisms used in the environment cannot be considered to meet the commonly understood meaning of "small quantities" for research and development, and thus do not qualify for the exemption.

d. *Implementation.* To implement the change in the R&D exemption, EPA intends to amend the PMN rule (40 CFR 720.3(cc) and 720.36) and possibly the SNUR general provisions in 40 CFR Part 720. The amendments will specify when a microorganism is considered not to qualify for the R&D exemption, and will provide enforceable standards for that determination.

Until the R&D rule amendments are final, EPA expects commercial researchers intending to release new, living microorganisms and engineered pathogens into the environment to report their activities to the Agency as explained in the units on PMN and SNUR notification (Units III.C.1 and 2). In addition, EPA strongly encourages researchers, prior to the time of reporting, to maintain records regarding containment procedures used in their experiments. Researchers should use the definition of "environmental release" provided in Unit IV.D as a guide, ask EPA for further guidance if questions arise, and in general be as inclusive as possible in their estimation of what should be reported.

e. Noncommercial R&D

Noncommerical R&D is exempt from section 5 of TSCA under section 5(g) and would therefore be exempt from

PMN and SNUR requirements even under the proposed amendments. EPA has defined "noncommercial" for all chemical substances subject to TSCA section 5 in a final rule published in the Federal Register of April 22, 1986 (51 FR 15096). As a general guide, R&D done by a commercial company should be considered commercial, and purely academic R&D should be considered noncommercial. For more specific guidance, the reader should examine the definition of "noncommercial" in the final rule and the discussion of "noncommercial" in the proposed PMN rule revisions published in the Federal Register of December 27, 1984 (49 CFR 50208). Readers should also note that the NIH Recombinant DNA Advisory Committee (RAC) and USDA Agriculture Biotechnology Recombinant DNA Advisory Committee (ABRAC) have jurisdiction over many noncommercial R&D activities, specifically recombinant DNA experimentation at institutions that receive funds from NIH and USDA. Both of these committees encourage submission of experiments from other sources as well.

4. *General information reporting requirements* — a. *Overview.* EPA intends to collect general information prior to the environmental use of microorganisms that are subject to TSCA, but that are not the subject of premanufacture or significant new use notification requirements. EPA will gather such information by means of a section 8(a) reporting rule. The information EPA collects will primarily be used to monitor environmental uses of microorganisms, thus making the Agency aware of cases that may require special regulatory action under other TSCA authorities. It will also be used to help the Agency evaluate and modify the scope of its biotechnology programs over time.

b. *Section 8(a) background.* Section 8(a) of TSCA authorizes EPA to issue rules requiring manufacturers, importers and processors of specified chemical substances to submit information to the Agency. TSCA section 8(a)(2) authorizes the Agency to obtain a broad range of data, including information on chemical identity and structure, production, use, exposure, disposal, and health and environmental effects. Small manufacturers, importers, and processors, as defined by EPA, are exempt from section 8(a) reporting and recordkeeping requirements, with certain statutory exceptions.

c. *Rationale for section 8(a) rule.* As explained in the overview to the EPA portion of this notice, the biotechnology review procedures described in this notice are intended to focus on the current areas of highest priority based on considerations of risk and on determinations about what makes a microorganism "new." However, there is a relatively high degree of scientific uncertainty involved in establishing these priorities at this early stage in the development of the biotechnology industry. The Agency cannot say definitively that all the microorganisms and uses that are not at this time subject to notification requirements will never need to be regulated or should never be subject to notification requirements in the future.

EPA believes that TSCA section 8(a) is the best mechanism available for determining whether specific microorganisms or categories of microorganisms not subject to PMN or SNUR notice requirements may need to be regulated. The Agency must be aware of how microorganisms are being used in the environment to fulfill its responsibility to identify and prevent important or immediate hazards that might unexpectedly arise with specific uses. The section 8(a) reporting with also provide EPA with necessary information to assess whether its overall priorities with regard to biotechnology regulation have been, in fact, appropriately set and whether they should change over time. As was pointed out by many comments on the Agency's first proposed statement on biotechnology, flexibility and incorporation of new information should be major components of any regulatory scheme.

d. *Implementation* — (1) *Who will have to report under section 8(a)?* When promulgated, EPA intends for this rule to apply to manufacturers, importers, and processors of microorganisms that are subject to TSCA and to be released in the environment, but are not otherwise reviewed under the PMN and SNUR policies described earlier. In other words, general information will be required prior to environmental releases of all microorganisms tht are subject to TSCA and that are non-engineered pathogens, or that are intra-generic or naturally occurring non-pathogens.

Although the rule will apply in general to the above groups, small manufacturers, importers, and processors are usually exempt from section 8(a) reporting and recordkeeping requirements. EPA has established general exemption standards for small manufacturers (40 CFR Part 704). The Agency will consider whether these standards should be retained or altered in some way to reflect considerations particular to the biotechnology industry.

When EPA issues its notice of proposed rulemaking, the public will have an apportunity to comment on the question of who will have to report under the rule.

(2) *What information will have to be reported under section 8(a)?* EPA is in the process of considering exactly what information it will propose to require on microbial products and uses under the section 8(a) reporting rule. In deciding what information should be reported on microorganisms, EPA will consider what information is necessary for the Agency to assess the safety of planned environmental releases, to evaluate its biotechnology regulations over time, and to consider necessary and

appropriate improvements. The Agency will also consider the economic impact of special information and whether the information is generally "known to or reasonably ascertainable by" potential respondents to the rule.

5. *Reporting of information on substantial risks.* All manufacturers, processors, and distributors of microbial products subject to TSCA, including those involved in research and development, are reminded of their responsibility to notify EPA immediately of any new information which "reasonably supports the conclusion that such substance or mixture presents a substantial risk of injury to health or the environment" (TSCA section 8(e)).

Guidance on the section 8(e) requirement was published in the Federal Register of March 16, 1978 (43 FR 11110). Manufacturers, processors, and distributors will find that this policy statement provides general guidance on TSCA section 8(e) reporting, but it should not be considered exhaustive in terms of the types of information that would reasonably support a conclusion of substantial risk. Specifically with regard to microorganisms, the types of information that should be reported include but are not limited to (1) pathogenicity to humans, plants, animals, or microbes, (2) significant ability to displace other organisms in the intended use area, (3) significant potential to transfer genetic material to other organisms, and (4) any other significant potential to cause harm to human health or the environment.

Manufacturers, processors, and distributors should be vigilant and immediately report substantial risk information concerning microorganisms subject to TSCA.

6. *Exemptions from premanufacture notification requirements.* Section 5(h)(4) of TSCA allows EPA, by rule, to exempt from PMN requirements chemical substances

that it finds will not present unreasonable risks. EPA expects to use this authority, where appropriate, to reduce the burden of PMN reporting requirements.

In its December 84 notice (at page 50891), EPA asked for comment on the issue of whether certain microorganisms or categories of microorganisms should be exempt from PMN requirements under the authority of section 5(h)(4) of TSCA. Ten respondents stated that microorganisms used in closed systems should be exempt under the 5(h)(4) provision, although several specifically remarked that appropriate biological and physical containment conditions should first be determined and met. Others suggested modifications to this approach, such as expedited reviews or reduced information requirements rather than outright exemption, or application of the exemption only to specific microorganisms or substances (e.g., *E. coli.* used in contained systems.). One commenter stated that an exemption was not appropriate because there is no current Federal authority to determine safety in the event of accidental release.

Under TSCA, the PMN policy described in Unit III.C.1 extends to commercial-scale, closed system used of microorganisms as well as environmental releases. The statute requires that all manufacturers of "new" substances must submit PMNs, regardless of whether they are used in contained facilities or open environments. Nonetheless, EPA believes that closed-system uses of new microorganisms will often present lower risks than environmental releases of the same organisms. The contained uses may therefore warrant a section 5(h)(4) exemption, and EPA is hereby announcing its intent to pursue that possibility.

Since the Agency does not yet have sufficient information to make the necessary finding under section 5(h)(4) that such activities "will not present an unreasonable risk of injury to human health or the environment," it is soliciting more data to support that finding in the case of closed system uses. The Agency would appreciate receiving data that would support an exemption either for all inter-generic microorganisms used in closed systems, or for specific categories of such microbes. For example, a category that has been suggested for exemption is inter-generic combinations involving microorganisms that exchange DNA by known physiologic processes, and that are on the NIH RAC exchanger list. This possible exclusion is mentioned in the OSTP preamble published in this Federal Register.

Information and data relevant to this issue should be sent to EPA at the address listed at the beginning of this notice.

In addition to supporting the use of section 5(h)(4) exemptions, the Agency will try to identify categories of microorganisms that pose lower risk even though they may not meet the necessary findings for exemption. In such cases, the Agency will consider reducing the burden of PMN reporting by lowering the information requirements associated with the PMN, and by conducting expedited reviews. The Agency requests any data or information that could be used to support exemptions or expedited reviews.

Bibliography

Callahan, D., *What Price Better Health: The Hazards of the Research Imperative.*

Clark, J., "Squandering Our Technological Future," *The New York Times*, August 30, 2001, p. A19. See also the presentation of Thomas Okarma, President and CEO of Geron Corporation, before the Council on September 4, 2003.

Conference of Catholic Bishops, in a news release ("Catholic Bishops Criticize Bush Policy on Embryo Research," made available by the United States Conference of Catholic Bishops, August 9, 2001).

Englund, U., *et al.*, "Transplantation of human neural progenitor cells into the neonatal rat brain: extensive migration and differentiation with longdistance axonal projections," Experimental Neurology 173: 1-21 (2002).

Faden, *et al.*, "Public Stem Cell Banks: Considerations of Justice in Stem Cell Research and Therapy" *Hastings Center Report* 33:6 (2003).

Faraci, M., *et al.*, "Severe neurologic complications after hematopoietic stem cell transplantation in children," Neurology 59: 1895-1904 (2002).

Kerr, D. A., *et al.*, "Human Embryonic Germ Cell Derivatives Facilitate Motor Recovery of Rats with Diffuse Motor Neuron Injury," The Journal of Neuroscience 23: 5131-5140 (2003).

Latham, S., "Ethics and Politics," *American Journal of Bioethics* 2(1): 46 (2002).

Lumelsky, N., *et al.*, "Differentiation of embryonic stem cells to insulin-secreting structures similar to pancreatic islets," Science 292: 1389-1394 (2001).

One useful description of this approach to the issue is Zoloth, L., "Jordan's Banks, A View from the First Years of Human Embryonic Stem Cell Research," *American Journal of Bioethics* 2(1): 7 (2002).

Onyango, P., *et al.*, "Monoallelic expression and methylation of imprinted genes in human and mouse embryonic germ cell lineages," Proceedings of the National Academy of Sciences of the *United States of America* 99: 10599-10604 (2002).

Orr, R., "The Moral Status of the Embryonal Stem Cell: Inherent or Imputed?" *American Journal of Bioethics* 2(1): 57-59 (2002).

Sato, N., *et al.*, "Molecular signature of human embryonic stem cells and its comparison with the mouse," Developmental Biology 260: 404-413 (2003).

Soria, B., *et al.*, "Insulin-secreting cells derived from embryonic stem cells normalize glycemia in streptozotocin-induced diabetic mice," Diabetes 49: 157-162 (2000)

Xu, R. H., *et al.*, "BMP4 initiates human embryonic cell differentiation to trophoblast," Nature Biotechnology 20: 1261-1264 (2002).

Zalzman, M., *et al.*, "Reversal of hyperglycemia in mice using human expandable insulin-producing cells differentiated from fetal liver cells," Proceedings of the National Academy of Sciences of the *United States of America 100: 7253-7258 (2003).*

Index

G

H

J

K

L

M

N

O

P

R